3D Printing with SketchUp

Use SketchUp to generate print-ready models and transform
your project from concept to reality

Aaron Dietzen

<packt>

BIRMINGHAM—MUMBAI

3D Printing with SketchUp

Group Product Manager: Rohit Rajkumar

Publishing Product Manager: Kaustubh Manglurkar

Senior Content Development Editor: Rashi Dubey

Technical Editor: Saurabh Kadave

Copy Editor: Safi s Editing

Project Coordinator: Sonam Pandey

Proofreader: Safi s Editing

Indexer: Hemangini Bari

Production Designer: Shankar Kalbhor

Marketing Coordinator: Nivedita Pandey

First published: May 2014

Second edition: February 2023

Production reference: 1060223

Published by Packt Publishing Ltd.

Livery Place

35 Livery Street

Birmingham

B3 2PB, UK.

ISBN 978-1-80323-735-0

www.packtpub.com

This book is dedicated to my amazing parents. To my late mother, who always encouraged my creative pursuits, and my dad, who bought me my first 3D printer. I would not be who I am without you both.

– Aaron Dietzen

Contributors

About the author

Aaron Dietzen has been modeling in 3D since high school when he started at his father's software company. Over the next few decades, he worked in technical support, training, and product management, learning about dozens of modeling and CAD software packages along the way.

One of the software packages Aaron learned about was SketchUp, which he taught himself to use. He currently works for Trimble, creating video content about SketchUp. Often referred to as the "SketchUp Guy," Aaron focuses on helping SketchUp users to develop their own workflows and become more efficient modelers, doing his best to make learning enjoyable. Aaron lives in Colorado with his wife, three kids, and a couple of rescued puppies.

I want to thank everyone I have connected with in the SketchUp community, and the members of the Trimble staff and users I have met, both virtually and in person. You guys are all amazing!

About the reviewers

Duane Kemp, a 3D reconstruction specialist living in Switzerland, started with Google SketchUp 6 in 2007 and has used every version since. His company Kemp Productions' animations were presented at *SketchUp Basecamp 2012*, and his work has been featured in books and articles. In the last few years, his company's client work has been in the field of non-proliferation.

He is the admin for Facebook's *Trimble SketchUp Group* and its LinkedIn version, *SketchUp User Group*, as well as other 3D and SU plugin groups. He has promoted the use of SketchUp since 2012 on social media platforms, continues to test plugins/rendering software, and absolutely enjoys using SketchUp.

Duane was the technical reviewer for Aaron's book *Taking SketchUp Pro to the Next Level*.

Aaron's 3D Printing with SketchUp *is a clear, easy-to-follow, and step-by-step instructive work that will get you up and running for successful 3D printing using SketchUp in no time.*

Sila Kayo is an experienced software developer with a focus on 3D, mobile, embedded, and server software development. He has used SketchUp to create models for a 3D flight simulator released on PC, mobile, and AR platforms. He is the founder of Kayo Games, which specializes in 3D simulations for PC, mobile, AR, and VR platforms. Currently, he lives in Oulu, Finland.

Table of Contents

6

Modeling Using Solid Tools   81

7

Importing and Modifying Existing 3D Models   101

8

Assembling the Pieces Post-Printing   119

Index   137

Other Books You May Enjoy   144

Preface

3D printing is an amazing technology and printing hardware is easier to own than ever before. While there are many other resources that cover the general process of 3D printing, this book is the ultimate guide to creating models for 3D printing from SketchUp.

Once we have established a basic understanding of how SketchUp is used in the 3D printing workflow, we will go through the steps to create a print-ready model using only SketchUp. Following that, we will cover how to use SketchUp to modify existing 3D files, and discover some additional tools that can be used to make SketchUp an even more powerful modeling tool. You'll learn how to transform 2D images into 3D printable solids, how to create multi-part prints that can be assembled without the use of fasteners or glue, and how to make sure your model, whether designed from scratch or assembled from pre-existing geometry, is ready to be made real via your own 3D printer.

By the end of this book, you should be able to generate 3D print-ready models from SketchUp.

Who this book is for

If you own or have access to a 3D printer and are tired of downloading and printing other people's 3D models, this book is for you! Learn how to use SketchUp to create your own custom pieces or modify existing files so you can print exactly what you need. Whether you are an architect hoping to print buildings, a designer needing quick physical prototypes, or a hobbyist wanting to print a tchotchke, this book is for you. You should have completed some training in fundamentals in SketchUp and be able to create and edit basic geometry.

What this book covers

Chapter 1, *Getting Started with 3D Printing and SketchUp*, is where we start with the basics and talk about how SketchUp fits in the 3D printing workflow.

Chapter 2, *Setting Up Your SketchUp Environment for 3D Printing*, discusses the different versions of SketchUp that are available and which you should use and dives into getting SketchUp ready to model for 3D printing.

Chapter 3, *Importing and Exporting .stl Files*, talks about getting files into and out of SketchUp.

Chapter 4, *Print-Ready Modeling and Scaling for Export*, explains what print-ready means and what scale you should be modeling at for 3D printing.

Chapter 5, Modeling from Scratch Using Native Editing Tools, uses SketchUp for Web to create some print-ready geometry using only native tools.

Chapter 6, Modeling Using Solid Tools, discusses what Solid Tools are and how to use them to create models for 3D printing.

Chapter 7, Importing and Modifying Existing 3D Models, explains the best practices for importing models made by others and how to work with them in SketchUp.

Chapter 8, Assembling the Pieces Post-Printing, discusses how to model multi-part prints that can be assembled temporarily or permanently.

To get the most out of this book

This is not a beginner's book. To get the most out of this book, you should complete some sort of fundamental instruction for SketchUp (I recommend checking out the learning tools at `https://learn.sketchup.com`). Additionally, since there are so many steps and options when it comes to creating and printing 3D models, the focus of this book will be on using SketchUp to create and edit geometry and we will not spend time looking at specific information about 3D printers, slicing software, or other 3D modeling programs.

Software/hardware covered in the book	Operating system requirements
SketchUp	Windows or macOS

The files are available through SketchUp's 3D Warehouse and are linked to the relevant chapter. While we will not go into detail, there are images shown throughout the book of slicing software, printers, and final prints. I used ChituBox and Cura to slice all the prints in this book. The printers I used were both from Elegoo (a Neptune 3 for FDM and a Mars 3 for SLA).

Conventions used

There are a number of text conventions used throughout this book.

`Code in text`: Indicates code words in text, database table names, folder names, filenames, file extensions, pathnames, dummy URLs, user input, and Twitter handles. Here is an example: "Once I completed my model in SketchUp, I exported it as an `.stl` file."

Bold: Indicates a new term, an important word, or words that you see onscreen. For instance, words in menus or dialog boxes appear in **bold**. Here is an example: "The **Solid Inspector** panel provides you with information and tools that will make it much easier to create solid models and repair issues that prevent your creations from being solids."

> Tips or Important notes
> Appear like this.

Get in touch

Feedback from our readers is always welcome.

General feedback: If you have questions about any aspect of this book, email us at `customercare@packtpub.com` and mention the book title in the subject of your message.

Errata: Although we have taken every care to ensure the accuracy of our content, mistakes do happen. If you have found a mistake in this book, we would be grateful if you would report this to us. Please visit `www.packtpub.com/support/errata` and fill in the form.

Piracy: If you come across any illegal copies of our works in any form on the internet, we would be grateful if you would provide us with the location address or website name. Please contact us at `copyright@packt.com` with a link to the material.

If you are interested in becoming an author: If there is a topic that you have expertise in and you are interested in either writing or contributing to a book, please visit `authors.packtpub.com`.

Download a free PDF copy of this book

Thanks for purchasing this book!

Do you like to read on the go but are unable to carry your print books everywhere? Is your eBook purchase not compatible with the device of your choice?

Don't worry, now with every Packt book you get a DRM-free PDF version of that book at no cost.

Read anywhere, any place, on any device. Search, copy, and paste code from your favorite technical books directly into your application.

The perks don't stop there, you can get exclusive access to discounts, newsletters, and great free content in your inbox daily

Follow these simple steps to get the benefits:

1. Scan the QR code or visit the link below

https://packt.link/free-ebook/9781803237350

2. Submit your proof of purchase

3. That's it! We'll send your free PDF and other benefits to your email directly

Part 1:
Getting Prepared to Print

3D printing is a vast topic with all sorts of opportunities to talk about workflows and tips. This first part of the book focuses on general information as to how SketchUp fits into this topic. While there will be some hands-on work, the first part really focuses on learning about what a 3D printing workflow with SketchUp looks like.

In this part, the following chapters are included:

- *Chapter 1, Getting Started with 3D Printing and SketchUp*
- *Chapter 2, Setting Up Your SketchUp Environment for 3D Printing*
- *Chapter 3, Importing and Exporting .stl Files*
- *Chapter 4, Print-Ready Modeling and Scaling for Export*

1

Getting Started with 3D Printing and SketchUp

3D printing is an amazing technology that allows you to take things directly from your imagination and make them into real items that you can hold in your hand. 3D printing has been around long enough that I believe we can skip over the history of the subject or listing out the reasons you might want to print something. In this book, we will assume that you have access to a **3D printer** and have a reason to print. Alongside that, based on the title of this book, we will also assume that you want to use **SketchUp** to get your 3D print created.

There are a few steps between imagining an item and finally making it exist in the real world, however. In this first chapter, we will look at the multiple steps involved in creating a successful 3D print and where SketchUp plays a part in this process. Once we've covered the basic workflow, we will dive into a few details and see some examples of SketchUp being used to create or prepare various files for 3D printing. Once you've completed this chapter, you will know exactly where SketchUp can be used as a useful tool in your 3D printing software tool kit!

In this chapter, we will cover these main topics:

- Understanding where SketchUp fits in the 3D printing workflow
- Knowing what SketchUp can produce
- Using SketchUp as a creation tool
- Using SketchUp as an editor
- Using SketchUp as a repair tool

Technical requirements

This chapter will be an overview of the 3D printing workflow and will not require any software or hardware. However, it will require a basic understanding of both SketchUp and the general concepts of 3D printing.

Understanding where SketchUp fits in the 3D printing workflow

3D printing is not a simple process of just pushing the "Print" button on your 3D printer and sitting back and waiting for the print to finish. To get your printer started on a print, you first have to create the geometry that will make up that print. This file then has to be made into a format that your printer will understand. Telling your printer to start printing the file is actually one of the final steps in the 3D printing workflow.

To understand where SketchUp fits into this workflow, let's take a look at the whole thing, step by step, from idea to final print. I think that the easiest way to visualize this workflow is shown in *Figure 1.1*:

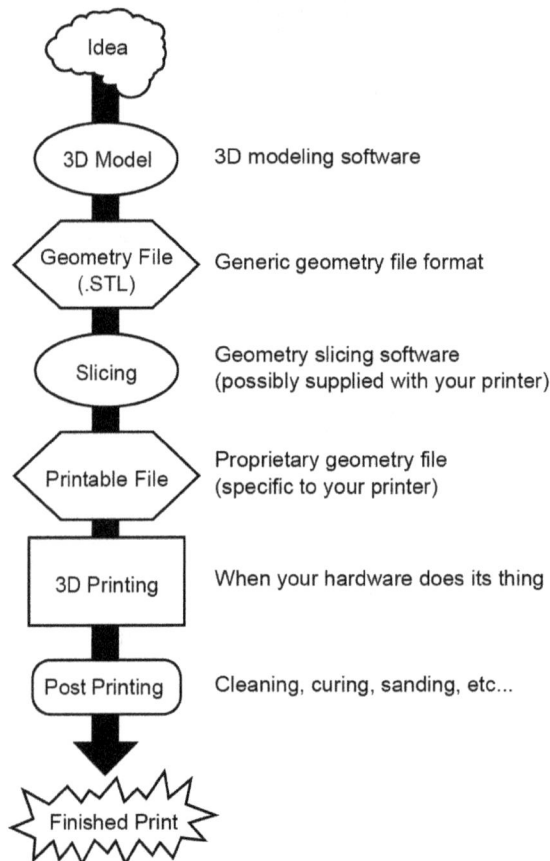

Figure 1.1 – Brief 3D printing workflow overview

While this is a general overview and may miss a specific step or two from your final workflow, this does cover the basics. To see exactly where SketchUp fits into this workflow, let me walk you through the process with an actual 3D print project:

1. **Idea** – This is where it all starts. Whether it is the need to replace a broken knob on your stove, create a piece for a board game, or bring a piece of art into the world, the first step is identifying what you want to create. I decided that I wanted to create a tiny coffee cup. Often, in the afternoon, my wife will ask for a cup of coffee. Since she does not want to be kept up at night, she will ask if I can make her a cup of "baby caffeine" coffee, which is made with mostly decaf coffee grounds and a little bit of regular. Next time she asks for this, I will pour her some coffee in a baby coffee cup!

Figure 1.2 – A hand drawing of my idea

2. **3D Model** – The next step is turning your idea into a 3D model. In some cases, this may mean finding an existing model to use as a starting point, while in other cases, it may mean starting from scratch. Regardless, you will need to create a three-dimensional representation of the thing that you want to print. This step is where SketchUp comes into play. In SketchUp, I can create a 3D model of the baby coffee cup. Any details I want in this print are included in the model and the final model will be at the proper scale.

Figure 1.3 – A 3D model of my baby coffee cup

3. **Geometry file** – While there are multiple file formats that can contain 3D geometry, most 3D prints end up, sooner or later, as a `.stl` file. This type of file contains simple information about 3D geometry, breaking it down into a bunch of connected triangles and showing where they exist in relation to each other. Once I completed my model in SketchUp, I exported it as a `.stl` file. In *Figure 1.4*, you can see that saving my coffee cup as a `.stl` file revealed all the triangular surfaces used to make up the smooth surfaces:

Figure 1.4 – The original SketchUp model on the left, the .stl file geometry on the right

4. **Slicing** – Now that the geometry is in a format that any program can read, it needs to be broken down into layers of 2D geometry that the printer will be able to create. Whether this file will end up being printed on an **FDM** or **SLA** printer, it will need to be sliced so that the printer can create the model one layer at a time. For my print, I used *ChituBox* to prepare my model (angling the model from the build plate, adding supports, and slicing for the printer). In the following figure, you can see the file that will be sent to the printer:

Figure 1.5 – My baby coffee cup design, sliced and supported

5. **Printable file** – Once the slicing software is done, it will export a file that can be read by the 3D printer. This file may be in a proprietary format (specific to the printer). Often, 3D printer users will use the slicing software that came with the printer, while others may use third-party software. This is largely a personal choice. The important part is that you end up with the files that will tell the printer what to print. In my case, as I am printing to an *Elegoo* printer, which reads *ChituBox* files, I exported a `.ctb` file format onto a USB drive.

6. **3D Printing** – Once the file is exported, it needs to find its way to the printer. This may happen via a USB cable, a flash drive, or over Wi-Fi. Once the file is on the printer, the magic happens. Well, it is actually less "magic" and more "waiting to see if the printer will print the whole thing without hitting any issues." This step is the part that you see in all the YouTube videos about 3D printing. This is where you get to see your idea come to life! With this step, my baby coffee cup has come to life!

Figure 1.6 – The baby coffee cup exists!

7. **Post printing** – I mention this step only because I am a realist. 3D printers are great in their ability to create real geometry from nothing. Fused plastic or cured resin turns our ideas into actual matter, which is awesome. Before you are done, however, you do have to clip support material, sand, and polish. There may even be some curing and painting before your vision is finally realized. For my project, this meant removing the supports, washing the cup in alcohol, curing the model in UV, then some light sanding and painting.

Figure 1.7 – This is my baby coffee cup, ready for paint

8. **Finished Print** – Eight simple steps later, you are holding your idea in your hand! My baby coffee cup is ready for my wife's next order of a "baby caffeine" afternoon coffee!

Figure 1.8 – The final print, painted and ready for use

When you step back and look at the steps, you will realize that there are quite a few of them. SketchUp will help you with two specific and very important steps. While all of these steps are necessary, the one that we will be focusing on in this book is the creation of 3D geometry and export of the `.stl` file.

Now that we are familiar with the broad strokes, let's dive deeper into the creation of a brand-new 3D model and a `.stl` file in the next section.

Knowing what SketchUp can produce

SketchUp is well known for being great software for creating 3D models. Users enjoy how easy it is to learn and how quickly they can generate 3D models of their ideas. SketchUp is used in dozens of industries, from architecture to product design, and has a large group of 3D printing enthusiasts using it for professional and hobby printing.

The most important point to understand is exactly what SketchUp can and cannot create. As we discussed in the *Understanding where SketchUp fits in* section, the 3D printing workflow is a process that includes multiple software packages and file formats. While it would be great if SketchUp could be the all-encompassing software, allowing you to go from an idea to a 3D print in one place, it just is not possible. Since different printers require different proprietary file formats, and users want different capabilities in their slicing software, it is impossible to cover the entire workflow with one software.

Despite the need for multiple pieces of software, SketchUp takes the most important step in this workflow in the creation of the 3D model. Based on the 3D printing hardware you own or have access to, you may be required to use one specific slicing software or to choose a specific software based on your own experience. Regardless, you can use SketchUp to create the initial 3D model and export the geometry file (`.stl`) that will be used by the slicing software.

There are also other ways to create 3D models for printing. In fact, there are probably dozens of software packages on the market right now that can create a 3D model and export it as a `.stl` file. All of them have strengths and weaknesses, and many excel at the creation of one type of geometry or another. Since this book is about 3D printing with SketchUp, let's focus on the advantages of using SketchUp in our 3D printing workflow and the types of modeling best suited to SketchUp.

Advantages of using SketchUp

Let's start by looking at the attributes of SketchUp that make it a great solution for creating 3D print-ready models:

- *Easy to learn* – SketchUp is very intuitive software. When you consider that most of the tens of millions of SketchUp users around the world are self-taught, you realize how quickly one can learn to use SketchUp.

- *Powerful* – Despite being easy to learn, SketchUp is extremely powerful. Not only is it a favorite among hobbyists, it is also a tool used in professional settings across the globe.

- *Multiple versions* – SketchUp offers multiple versions for different uses and budgets. SketchUp Free, SketchUp Go, or SketchUp Pro can be used to create 3D models that can be used for 3D printing. Whatever your budget is for software, there is a SketchUp version that should work for you.

- *.Stl support* – SketchUp natively supports the creation and editing of `.stl` files. This is the primary file format used to pass geometry around in the world of 3D printing.

- *Multiple file formats* – In addition to `.stl` files, SketchUp supports a dozen of other file formats. While not all of these are regularly used to import or export 3D geometry, many of the common formats (such as `.dxf`, `.obj`, or `.dae` files) are available.

There are quite a few options out there when it comes to 3D modeling software that can be used to create a file for 3D printing. When given this list, however, you can see why so many people end up using SketchUp as a part of their 3D printing workflow!

Model types ideal for SketchUp

While SketchUp is a commonly used tool in the realm of modeling for 3D printing, it is not perfect for everything that you might want to 3D print. Like other 3D modeling software, there are types of geometry that it excels at. SketchUp works best when modeling these types of geometry:

- **Hard-sided modeling** – SketchUp was originally created to be an architectural modeling software. As a result, many of the native tools and capabilities work great for creating hard-sided geometry (as opposed to more organic modeling). Things such as models of buildings or parts defined by distinct geometric shapes are easy to model in SketchUp.

  This does not mean that organic or flowing shapes cannot be created. SketchUp can absolutely make these sorts of shapes but, in some instances, may require additional software in the form of an extension to do so.

- **Scaled models** – SketchUp has great tools in place that will allow you to work on a model at two scales. This is a great option to have if you are modeling something that exists in the real world but will need to be printed on a smaller scale.

  For example, if you wanted to model the house you live in and print out a version that could sit on your desk, you would want to model it using dimensions that you could take from the real house. You could then use SketchUp to scale the model down so that it is only 4" long. The nice part is that scaling does not have to be done after you have finished modeling full size, but can instead be part of the modeling process from the start.

- **Multi-part models** – Another thing that SketchUp allows you to do in real time is to break your model into smaller pieces while you are modeling. If the model you are printing is bigger than what will fit on your printer, you can use SketchUp to break it into smaller pieces and include mechanical connections between the two pieces in the working model.

Hopefully, you can see that SketchUp is a great tool for working with 3D geometry for printing. Now let's dive into the different ways you can use SketchUp to create print-ready models.

Using SketchUp as a creation tool

My primary use of SketchUp for 3D printing has been as a tool to create geometry for printing from scratch. I have created many prints from SketchUp models, and most of the time I was working from an idea or basic drawing to create the model that ended up being printed. In these cases, I get to use SketchUp starting from an empty file and start modeling the geometry that I want to print, as I see it in my mind.

As an example, let me show you my haunted house model. This model was created as a prop for a children's book (I never got around to finishing) and was modeled completely from my imagination:

Figure 1.9 – My haunted house model in SketchUp

When I created this, I was able to start with basic shapes in SketchUp (literally a bunch of boxes) that I could push around until I found the right shape for the house. Once the general layout was created, I broke the whole thing down into pieces.

Since I knew that I wanted to print this fairly big (around 6" wide), I knew that I could not print it as a single piece. The printer I was using at the time (an *Elegoo Mars*) had a build volume of 4.7" x 2.6" x 6.1". This was limiting, but I knew that I could break this house down into pieces that would fit.

Additionally, I wanted to print this house hollow. In the end, it would be large enough that I would need an entire bottle of resin if I were to print it solid. To achieve all of this, I broke up the model as if it were a model kit.

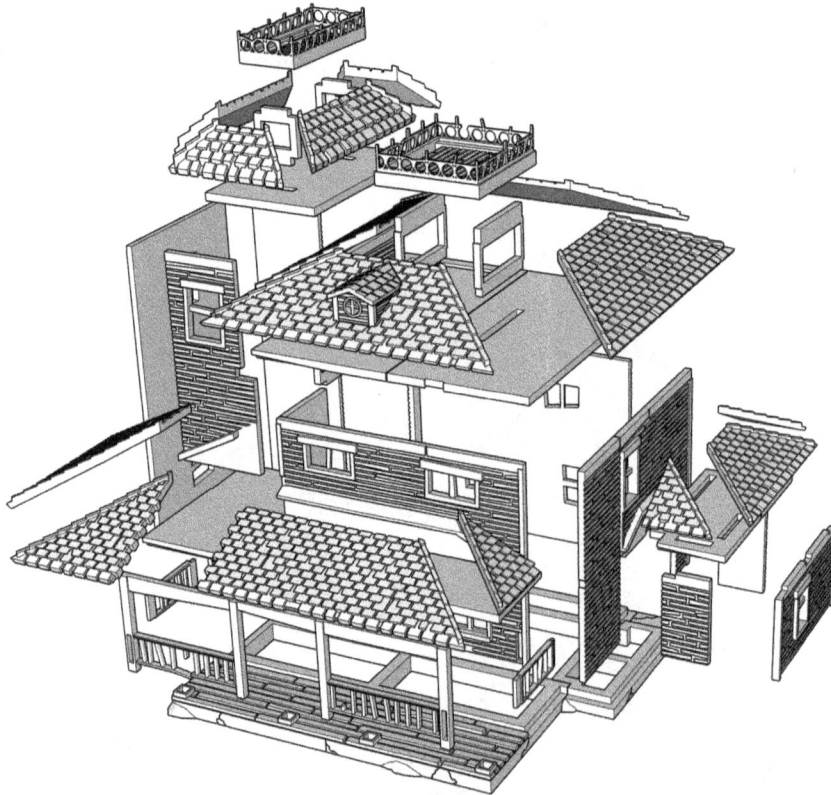

Figure 1.10 – Exploded view of the haunted house pieces

These pieces could be printed, in some cases multiple pieces in a single run, and glued together. There was some learning on my part regarding how to best orient these rather thin and long chunks of geometry so that they would print well. In the end, I got my process dialed and was able to get all the pieces printed and assembled.

Figure 1.11 – The final print, assembled and painted

I allowed myself to create something that I had no actual reference for, other than an image in my head. Additionally, with SketchUp I was able to break the house down into printable pieces and experiment with geometry, until I found the ideal pieces to print.

While working from scratch is great, there are times that you may be working from an existing geometry. Fortunately, SketchUp makes that easy to do as well, as we will see in the next section.

Using SketchUp as an editor

One of the most amazing things about the 3D printing community is the members' willingness to share their printable creations. In some cases, you may purchase models from a creator or a company, or even use online tools to generate models of things, such as characters for games. In other cases, you may be able to download shared models for free and print them right away.

Depending on the model you download and your specific use for the print, you may want to make a change to the geometry before printing. This may be as simple as changing the size of the model but may be more involved, such as adding a figure to a custom base, or adding or removing geometry so that the model can serve a specific purpose in the real world.

> **Resizing models in the slicer**
>
> Every single slicing program I have used has had the option to set the scale of the imported geometry. This works just fine if you want to make it a little bigger or a little smaller, or if you need to scale to get your model to fit on the build plate. Where this can fall short, however, is if you want the model to be a specific height or length.

Take, for example, this ghost print. I printed a copy, but the bottom of the print did not sit flat. To fix this, I decided to add a base. I started by importing the `.stl` file in to SketchUp.

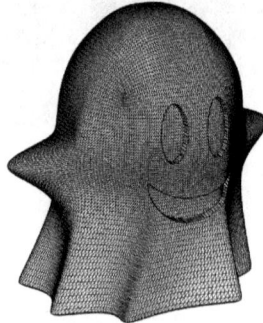

Figure 1.12 – Imported .stl ghost mesh

Once the mesh was imported, I modeled a quick circular base in SketchUp and merged the two pieces together. This created a new model that I could print that I know will sit flat on my desk.

Figure 1.13 – The imported ghost mesh merged with the new base geometry

Edits to imported models can be much more extensive and can include things such as breaking models into smaller, printable chunks or adding mechanical connection points between pieces. Basically, if you have a `.stl` file of a model and want to change anything about it, then you can make those changes in SketchUp.

Occasionally, you will have an issue with a model that just needs to be fixed, as opposed to any kind of edits. Let's talk about repairing geometry in the next section.

Using SketchUp as a repair tool

At times, you may come across a 3D model that your slicing software will not accept. While it would be great to think that a 3D model would be ready to print any time anyone created and shared it, this is just not always the case. It is not too hard to find examples of a `.stl` file that is missing faces from the mesh, or a 3D model that is completely the wrong size.

While it is easy to have a "plenty of fish in the sea" mentality and throw out these models and go look for alternatives, there are situations where you may have found the perfect file for your needs, but you just cannot get it to print. In that case, you may want to try importing the model into SketchUp and see if you can take care of the problems, manually.

> **Fixing it in the slicing program**
>
> It should be noted that many slicing programs out there have some repair options built into them. In some cases (such as a single missing face in the mesh), a fix might be as simple as clicking a button and letting the slicer fill in the gap. In many cases that I have seen, however, these repair tools are very basic and cannot fix anything but the most basic of issues.

I have this `.stl` file of a dog model that I want to print for a friend:

Figure 1.14 – A .stl file that looks good from a distance

The model looks good, but when I take it into my slicer it ends up looking weird. The problem is that there are holes in the mesh, and my slicer can't seem to be able to figure out how to fix the problem.

Using SketchUp, I can spin the model around quickly in 3D and zoom in on the mesh to see if I can identify any holes (I can even use a few different tools or extensions to help speed up the process, but we will get into them in *Chapter 7, Importing and Modifying Existing 3D Models*). In this specific example, if I orbit around the dog and take a look at the bottom of his front foot, I will see an issue.

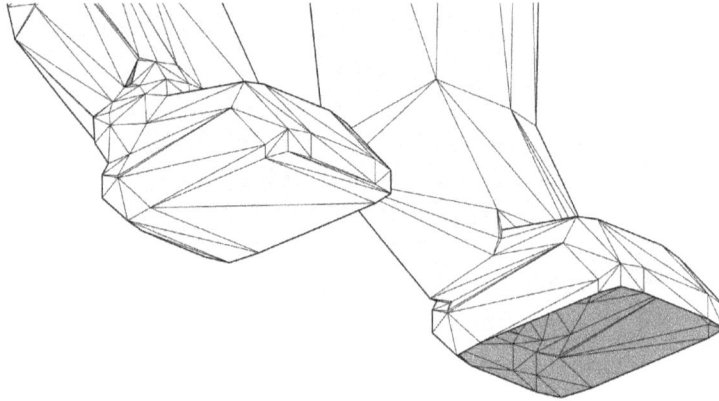

Figure 1.15 – Faces missing from the bottom of the dog's foot

This is very easy to fix but, once it's repaired, there is still an issue with this mesh. If I continue to explore the file, I find that the dog's other front foot is missing a tiny little triangle.

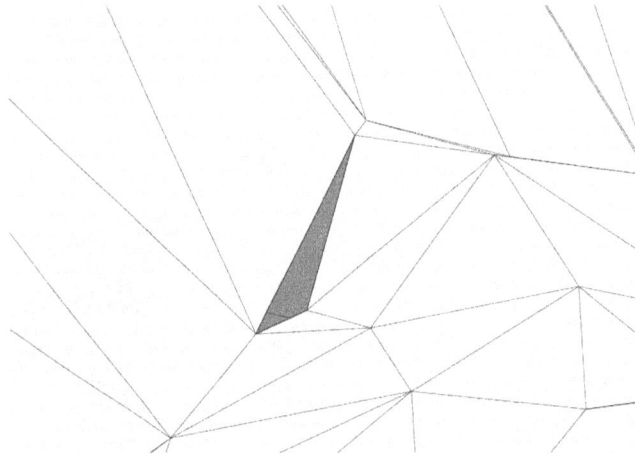

Figure 1.16 – One more missing face in the mesh

Once I use SketchUp to create some very basic geometry to fill the holes, I can export a new `.stl` file and send it off to my slicer to get it ready for printing.

Summary

Hopefully, this chapter gave you a good idea of what the entire 3D modeling and printing workflow looks like and where SketchUp fits in. You should now know what sort of modeling can be produced with SketchUp and have at least an idea of how it can be used to create models from scratch, edit existing models, and can be used as a repair tool.

I believe that the best way to learn to use a tool such as SketchUp is to use it, so we will spend quite a bit of time modeling and editing models in SketchUp. Before we do that, however, let's talk about which version of SketchUp you should use and some ways to get your SketchUp ready to create print-ready models in *Chapter 2, Setting Up Your SketchUp Environment for 3D Printing*.

2

Setting Up Your SketchUp Environment for 3D Printing

Nowadays, when it comes to selecting a program to use to create a 3D model for printing, we are spoilt for choice. There is both free and paid software that allows you to create 3D geometry in different ways using different modeling tools. Since you are reading this book, we will spend some time in this chapter exploring the advantages of using SketchUp to create 3D models for printing.

Personally, I think that SketchUp is an amazing choice of 3D modeling tool, and I say that having used a dozen different modeling software packages over the years. To get SketchUp to make exactly what you need when it comes to 3D modeling, you will need to spend a little bit of time upfront learning about how SketchUp does what it does and then getting it set up to start modeling. Fortunately, this is exactly what this chapter will help you to do!

In this chapter, we will cover these main topics:

- Understanding SketchUp's capabilities

- Deciding which version of SketchUp to use

- Getting SketchUp ready to make print-ready models

Technical requirements

To follow what is shown in this chapter, you will need your computer, an internet connection, and a subscription to SketchUp Free, SketchUp Go, or SketchUp Pro.

Understanding SketchUp's capabilities

As I mentioned in the introduction to this chapter, there are a lot of different 3D modeling software options out there and they have some very different approaches to creating 3D geometry. Some modeling software focuses on using tools to generate rigid, hard-sided geometry, while others have tools to allow you to sculpt geometry in 3D space similar to creating a clay model. There is even software that allows you to "paint" space in virtual reality, then export your 3D paintings to be printed in 3D!

When it comes to modeling in SketchUp, it can help to remember where SketchUp comes from and what the tools were created for initially. SketchUp was originally created to be used as an architectural modeling tool. For this reason, many of the modeling tools make sense when you think about modeling things such as buildings, as opposed to something small and soft, such as a rose, for example. When SketchUp was originally created, the tools were written with the intention of making large, building-sized models with rigid geometry and hard angles.

This is really more of something to keep in mind rather than a rule that you need to follow. When I look at the geometry that I have to create in 3D, I try to see three different levels:

- **Level 1 geometry** – This is geometry made up of basic shapes: things such as extruded rectangles, spheres, and other 3D volumes made of polygons. This does not mean simple. A model of the exterior of a full house is anything but simple but is still made of basic shapes.

- **Level 2 geometry** – This is geometry that has more organic shapes to it, such as the rose that I mentioned earlier or the basic shape of a tree. This is geometry that does not have identifiable polygons creating the shapes. Again, this is not about how large the shapes are or how many shapes are needed, but rather a recognition that a level 2 model includes some amount of organic geometry.

- **Level 3 geometry** – This is geometry that is a fully detailed 3D surface. This is a detailed model of a face or a full-body 3D figure with skin details. This is the sort of geometry regularly created in 3D sculpting software as opposed to polygon modeling software.

For those of you that prefer a visual example of these levels, here is the sort of geometry I am talking about:

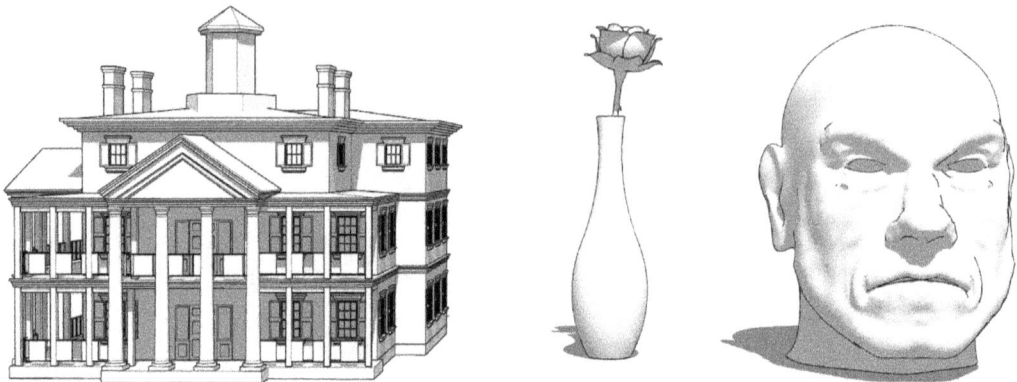

Figure 2.1 – Left to right, a level 1 house, a level 2 rose, and a level 3 human face

Looking at this list, I would use SketchUp to create level 1 geometry all day long. SketchUp's native toolset was created specifically to create this sort of geometry. Examples of these sorts of models would be a desktop-size version of the home I am designing or a custom elbow for my shop vac.

When it comes to level 2 geometry, I would not shy away from modeling this in SketchUp, but it may mean spending a little more time thinking about how to accomplish the model before diving in. It may also mean using SketchUp Pro and taking advantage of the wealth of extensions available. Examples of this sort of geometry might be something such as a knight chess piece or the model of a rose that I mentioned earlier.

Level 3 geometry is something that I would not plan on doing in SketchUp. If I needed a fully detailed bust of someone 3D printed, I would lean on software that excels at that sort of work.

I know that there are a few hardcore SketchUp modelers out there who will argue that you can – given enough time and energy – model anything in SketchUp, who would push back on me for saying this, but the fact is, there are multiple tools out there for a reason. Just as I would not want to try to model an accurate example of a steel framing detail to print out in miniature to sit on my desk using sculpting software, I would not want to use SketchUp to create a super detailed dragon.

My hope with this is not to tell you that you should not use SketchUp for 3D modeling, but quite the opposite. SketchUp is a fast and easy way to generate most of the print-ready models that I come across. I added this part of the chapter so that you understand that as we move through this book, we will not be spending time modeling detailed action figures using SketchUp because it is not the perfect tool for that job.

Now that we are all on the same page about what SketchUp can do, let's talk about which version of SketchUp is right for you.

Deciding which version of SketchUp to use

One of the great things about SketchUp is that it is available in several different versions. From the free version to a high-end package that includes things such as rendering software and some advanced construction tools, SketchUp has a version that works for most workflows. While we do not need to go too deep into the SketchUp offerings, I do want to take just a moment to look at the different ways that you can get SketchUp running and the functionality available in each version. Before we look at actually using SketchUp, let's cover the different ways you can get hold of SketchUp.

Applications versus plans

There was a day, not too long ago, when getting SketchUp on your computer was a pretty simple process. Today, it is not difficult, but you will need to think about the version that you want before you start downloading software or subscribing to anything. For this reason, I want to spend just a few minutes explaining how SketchUp is licensed, the difference between SketchUp plans, and the applications that the plans include.

There are four plans through which you can run SketchUp: SketchUp Free, SketchUp Go, SketchUp Pro, and SketchUp Studio. Each of these plans includes a version of SketchUp as well as additional software that can be used alongside SketchUp in various workflows.

> **No SketchUp Studio**
>
> Since we are talking about using SketchUp for 3D printing, I am not going to review the contents of SketchUp Studio as it contains everything in SketchUp Pro, plus additional tools. Since these tools bring nothing to the 3D printing workflow, I don't see a good reason to go into detail about what it includes.

Each of these plans includes a version of SketchUp that can be used to model 3D printable geometry. Some versions make this easier than others with added functionality, which is exactly what we will investigate right now.

SketchUp Free

SketchUp Free is a version of SketchUp that is, as the name suggests, free to use. The one caveat with it being free to use is that it cannot be used for commercial work. If you sign up for a free account, you have to agree to an **End User License Agreement** (EULA) that says as much. As a 3D printer, this means that the work created in this version would be for your own use or enjoyment. If you are printing models as a part of your job or plan to print and sell models, then you will need to use to SketchUp Go or SketchUp Pro.

SketchUp Free does not have any software to download or install. With SketchUp Free, you have access exclusively to SketchUp for Web. As with any version of SketchUp, SketchUp for Web is perfectly capable of creating 3D models that you can export and send to your 3D printer. The modeling tools you need to create 3D geometry are the same across all versions of SketchUp and can be used to create 3D printable solids in SketchUp for Web in the same way as in SketchUp for Desktop.

SketchUp for Web also has basic file export, which includes .stl export. This means that, once you have completed your 3D model, you can download the .stl file to send to your slicing software. The SketchUp Free version of SketchUp for Web does not, however, have the ability to import .stl files. If you have the need to work with existing .stl files (something we will be covering in detail in *Chapter 3, Importing and Exporting. stl Files*), then you will need to look into one of the other plans.

Additionally, there is some functionality missing from the free version of SketchUp for Web that you may want as a part of your 3D print modeling workflow, namely **Solid Inspector** and **Solid Tools** (we will see both of these and the impact they make on modeling for 3D printing in *Chapter 6, Modeling Using Solid Tools*). These are tools that we will be using throughout the examples in this book. While in the free version of SketchUp for Web, it is possible to model solid geometry without these tools, they will make modeling for 3D printing much easier in the long run.

Let's end this section with a quick pros and cons list for SketchUp Free:

- SketchUp Free pros:
  - Free to use
  - Includes SketchUp for Web
  - Native 3D modeling tools
  - No downloads or installations required
  - Can export .stl files
- SketchUp Free cons:
  - Cannot use for commercial work
  - Cannot import .stl files
  - Missing **Solid Inspector**
  - Missing **Solid Tools**
  - Cannot use extensions
  - You must be connected to the internet to use SketchUp

There are many different reasons to create models for 3D printing. As you look through this list, think about what is important to you and how your workflow might be impacted by the items on these lists.

SketchUp Go

SketchUp Go is the most affordable SketchUp plan. This plan includes an upgraded version of SketchUp for Web as well as access to SketchUp for iPad. In the Go version of SketchUp for Web, you have access to everything from the free version, plus the ability to import and export .stl files. You can also use the in-built **Solid Inspector**. This is a super useful tool that will check your model to make sure that it is solid (an exported .stl file needs to be solid in order for the slicer to properly do its job and get your model printed).

Finally, the Go version of SketchUp for Web includes **Solid Tools**, which allows you to use basic solid shapes to add, subtract, or merge to create single solid pieces. We will look at modeling with **Solid Tools** in-depth in *Chapter 6, Modeling Using Solid Tools*. However, if you are planning to do 3D prints, the ability to work with solids is key, and **Solid Tools** gives you the ability to work with solids in a way that will save you time and energy.

Additionally, there is no EULA limitation on who can use SketchUp or for what purpose. This means you can generate print-ready geometry for your job or to sell without any issues.

Yes, there is a small monetary commitment to using SketchUp Go over SketchUp Free, but may be worth it for what you get in this package. Plus, as far as the price goes, it is a fraction of what I would spend each month on filament or resin for my printers. If I think about it as a tool that I need to keep printing, the price seems pretty minor in the scheme of things.

Here is the list of the pros and cons of SketchUp Go:

- SketchUp Go pros:

  - Inexpensive
  - Includes an upgraded version of SketchUp for Web
  - Includes **Solid Tools**
  - Includes **Solid Inspector**
  - No downloads or installations required
  - Can import and export `.stl` files
  - Includes SketchUp for iPad

- SketchUp Go cons:

  - Cannot use extensions
  - You must be connected to the internet to use SketchUp

Once again, think about what you need as far as a modeling tool goes, and look at what you get with this plan. If you are planning on making money off 3D printing or will be creating prints for your job, this is the entry level for you and an amazing tool to have in your 3D printing software toolbox.

SketchUp for iPad

SketchUp for iPad is included in both SketchUp Go and SketchUp Pro, yet we are not covering it in this book. Why is that? SketchUp for iPad is an amazing program and very useful for modeling on the go or letting you model with an Apple pencil rather than a mouse. However, to use it, you do need an iPad, which was not an assumption I wanted to make. Additionally, 3D printing requires a lot of files passing back and forth between SketchUp, the slicing software, and the printer. Often, these files end up getting to the printers on flash drives, memory cards, or through a USB cable, and I did not want to spend time talking through different ways to get files off the iPad after you had modeled them. Suffice it to say, if you prefer modeling on SketchUp for iPad, a lot of what you can do in SketchUp for Web can also be done there, but you will need to work out your own method of transferring files.

SketchUp Pro

SketchUp Pro includes everything that is in SketchUp Go, plus it includes SketchUp for Desktop. SketchUp Pro includes many other features as well (such as LayOut, the XR Viernes, and access to PreDesign) but for the purpose of 3D printing, it is all about having SketchUp for Desktop!

SketchUp for Desktop allows you to download and install SketchUp on your PC or macOS and run it locally. With the software installed, you do not need to be connected to the internet to use SketchUp. Additionally, SketchUp for Desktop performs better than SketchUp for Web and will allow you to work with larger, more detailed files. One of the biggest differences, however, is the ability to use extensions.

Extensions are little programs, often written by third parties, that add to or modify the way that SketchUp for Desktop works. An extension can do something simple such as check to see whether a model is solid and ready to print, or a complicated process such as creating a realistically draping piece of cloth. Extensions allow you to "extend" the capabilities of SketchUp far beyond what it can do by itself.

SketchUp Pro is more expensive than SketchUp Go, but for many people, not prohibitively so. If you are 3D printing as a hobby, SketchUp Free or SketchUp Go might work for you, but if you are 3D printing as an income-producing activity, it is pretty easy to offset the price of SketchUp Pro (which is less than the cost of most printers), and worth the expense when you consider how much software you end up with!

Here is the list of the pros and cons of SketchUp Pro:

- SketchUp Pro pros:

  - Includes an upgraded version of SketchUp for Desktop

  - Includes **Solid Tools**

  - Includes **Solid Inspector** (as an extension)

  - Allows the use of other extensions

  - Installed software

  - Can import and export `.stl` files

  - Includes SketchUp for Web, SketchUp for iPad, LayOut, and more

- SketchUp Go cons:

  - Higher cost per year

If you are using SketchUp as a 3D modeling tool for work and doing some 3D printing on the side, or if you are selling 3D prints and making a profit, SketchUp Pro is probably the best plan for you.

This section was meant to serve as an overview of the options when it comes to which versions of SketchUp are available. As we move into specific, hands-on examples of using SketchUp in the next few chapters, we will look at using each of these and how they can be useful in the 3D printing workflow. Before we do that, though, let's take a look at how you would go about setting each one up for 3D printing in the next section.

Getting SketchUp ready to make print-ready models

Regardless of the version of SketchUp you end up using, you will want to spend a little bit of time getting ready to create print-ready models. Let's start by looking at what you need to do to get each version of SketchUp ready for modeling and what commands or tools will be the most important. Then, we will look at a few things that you may want to keep near you in the real world as you model.

Preparing SketchUp Free

If you decide that the free version of SketchUp for Web is adequate for your modeling needs, you are in luck, as there is actually very little you will need to set up to start modeling! Even though SketchUp Free is free, you will still need to create an account and log in. Plus, creating an account allows you to gain access to Trimble Connect, where you can store your models online as well as save some basic settings.

Once logged into the free version of SketchUp for Web, there is not a whole lot to set up, but there are a few things that you will want to be aware of as you think about creating models for 3D printing. In this chapter, we will briefly cover where some of the controls are that we will be referring to in later chapters, where we do some step-by-step modeling in SketchUp for Web.

The first is the units that you will be using to create your model. We will be getting into modeling in real-size increments versus exporting in smaller units in *Chapter 4, Print-Ready Modeling and Scaling for Export*, but it is worth knowing that units are set or changed in this version of SketchUp via the **MODEL INFO** panel on the right side of the screen.

The **MODEL INFO** panel is accessed by clicking on the second to last icon on the right side of the screen (a lowercase *i* in a circle), as seen in the following figure:

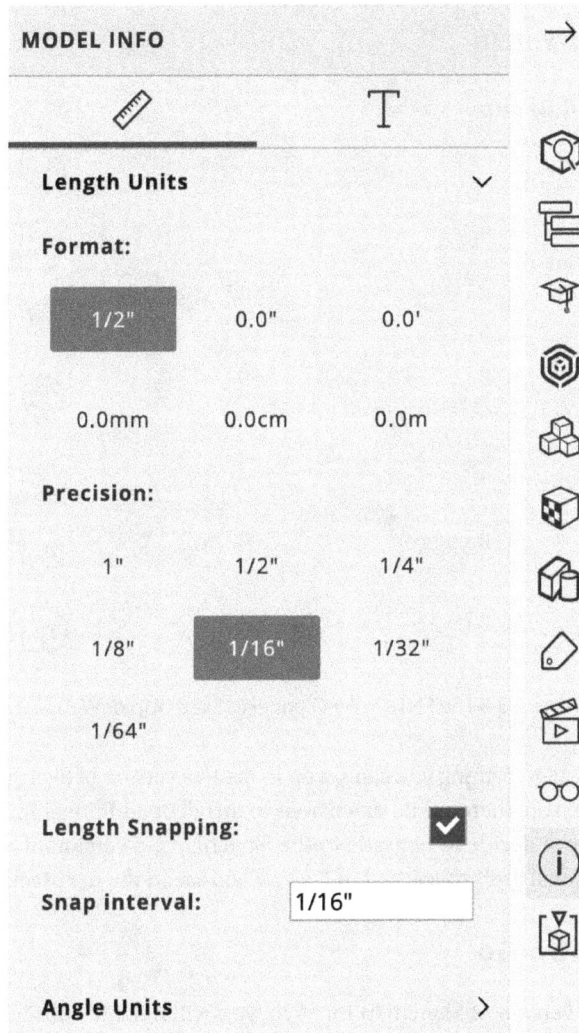

Figure 2.2 – The MODEL INFO panel from SketchUp for Web

The other panel we will use a lot in SketchUp for Web is the **ENTITY INFO** panel. This is the panel that will let us know whether the model we are working on is solid (thus print-ready) or not. Access the **ENTITY INFO** panel by clicking on the first icon on the right side of the screen (the cube with a magnifying glass), as seen in *Figure 2.3*:

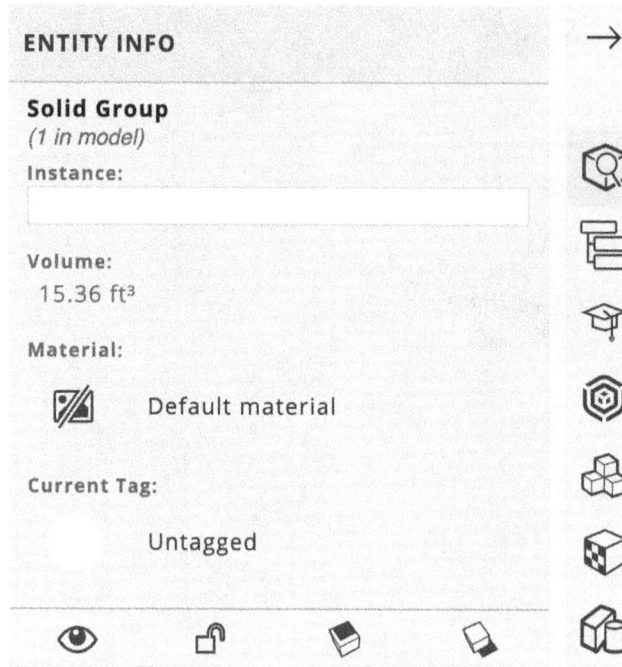

Figure 2.3 – The ENTITY INFO panel in SketchUp or Web

With that, we have looked at everything you can set up in the free version of SketchUp for Web. Unlike the other versions of SketchUp, there are no extensions to install or additional functionality of which you need to be aware. If you decide to upgrade to the SketchUp Go version of SketchUp, however, you will have some additional tools to learn about, as we will see in the next section.

Preparing SketchUp Go

To use the SketchUp Go version of SketchUp for Web, you will need to subscribe rather than just create an account. You can purchase a SketchUp Go subscription at the main SketchUp website (`https://sketchup.com`) or by clicking the **Upgrade** button in the lower-left corner of the free version of SketchUp for Web. If you have not already done so, you will need to create a Trimble ID before completing your purchase. Once you have successfully subscribed, you will be able to log in at `https://app.sketchup.com` and get access to the full version of SketchUp for Web.

In addition to the panels that we reviewed in the *Preparing SketchUp Free* section (if you skipped past this section and jumped right here, it may be worth it to run back and read it), there is another panel that you will want to use regularly as you model. This is the **Solid Inspector** panel. The **Solid Inspector** panel provides you with information and tools that will make it much easier to create solid models and repair issues that prevent your creations from being solids. It can be accessed by clicking on the last icon on the bottom-right side (the one that looks like a little cube inside of a 3D printer), as shown here:

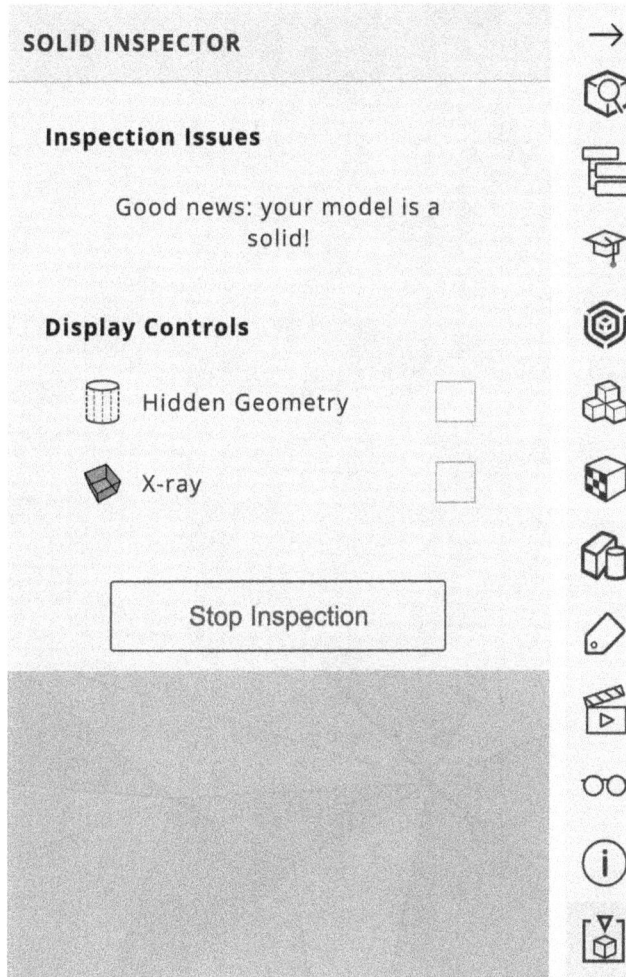

Figure 2.4 – Solid Inspector in the Go version of SketchUp for Web

The other tools that you get access to with SketchUp Go that are beneficial to modeling for 3D printing are **Solid Tools**. These tools are available in the toolbar on the left side of the screen and can be accessed by clicking on the icon with the two overlapping squares, as seen in *Figure 2.5*:

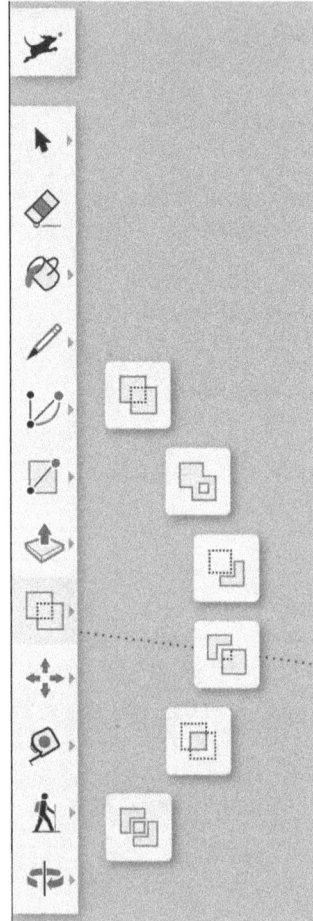

Figure 2.5 – Solid Tools in the Go version of SketchUp for Web

While these two bits of added functionality may not seem like a huge list of upgrades, we will see that they can be huge time savers as we create models for 3D printing (as we will see in *Chapter 6, Modeling Using Solid Tools*).

You have probably noticed that with SketchUp for Web, there is not a whole lot that needs to be set up beforehand. SketchUp for Desktop, on the other hand, has things such as templates and extensions that we will want to look into using before we start modeling.

Preparing SketchUp Pro

Unlike SketchUp Free or SketchUp Go, SketchUp Pro will allow you to download and install SketchUp for Desktop. Before modeling, you will have to get this installed and sign in using the Trimble ID and password associated with your subscription. Once logged in, you can get SketchUp for Desktop ready to create models for 3D printing. SketchUp Pro includes almost everything that is in SketchUp for Web, but with a few changes in the user interface.

Same tools, but in a different location for SketchUp Pro

When it comes down to the basics, creating 3D geometry in the different versions of SketchUp is very similar. Most of the drawing and editing tools are the same, though the user interface can be slightly different. When we look at SketchUp for Desktop, all of what we looked at in SketchUp for Web applies, but the location of the tools or information will be different.

Units

Unlike SketchUp for Web, SketchUp for Desktop keeps its **MODEL INFO** screen in the **Windows** menu, rather than a panel on the side. Additionally, the **MODEL INFO** window in SketchUp for Desktop includes a wealth of information and tools. Things such as default settings and information about your current model can be displayed in SketchUp for Desktop.

Entity Info

The **ENTITY INFO** panel is just as informative in SketchUp for Desktop as it is in SketchUp for Web. In SketchUp for Desktop, however, it is not always on the top of the right side of the screen. Since you have the ability to customize your user interface in SketchUp for Desktop, the location of this panel may change. Additionally, panels are displayed differently in Windows than in macOS. Regardless of your operating system, it will be important to make sure that your **ENTITY INFO** screen is available to you while you model.

If you are running SketchUp for Desktop on a PC, you can open **Default Tray** from the **Windows** menu. The first panel in **Default Tray** is **ENTITY INFO**.

If you are a macOS user, you can turn on the **ENTITY INFO** floating panel through the **Windows** menu.

Solid Tools

Unlike SketchUp for Web, where the **Solid Tools** commands are in the default toolbar, **Solid Tools** has its own toolbar in SketchUp for Desktop. This is not a huge change, but it is kind of nice to have control over the UI and have the **Solid Tools** toolbar on and located wherever you want it on your screen.

Features that are not in SketchUp for Web

Unlike SketchUp for Web, SketchUp for Desktop has a few options that make it more powerful. The most obvious is the fact that it is an installed application, rather than software running off a web page. While you may be able to generate the same sort of models in either version, with the Desktop version, you do not have to rely on always being connected to the internet, and the power of your computer will have an effect on how well SketchUp runs.

Additionally, SketchUp for Desktop allows you to use two powerful features that are not in SketchUp for Web: templates and extensions.

Templates

Templates are, simply put, SketchUp files that have whatever you want in them when you start a new model. With a template, you can set things such as the units you want to use, the initial view of your modeling space, and even the addition of things such as your print bed in the model.

For the purposes of modeling for 3D printing, our template needs are simple and may actually change from one model to the next. I end up using one of the two default templates in most of my modeling, choosing the template that uses the units that I want to model in. The preferred template depends on the item that I am modeling. If I am modeling a full-size item, such as a vehicle or a structure, I will model using the **Architectural** template in **Inches**. If I am modeling something that is at scale, such as a replacement part, I will measure and model it in millimeters, so I will use the **Woodworking** template in **Millimeters**.

While it is possible to create and save your own templates, specifically for 3D printing, I don't find there is a need. If, as you proceed through the examples in this book, you find that there are changes you make to your starting model every time, then you may want to save a custom template.

> ### Stock 3D printing templates
>
> If you are running SketchUp for Desktop, you may notice that there are a couple of 3D printing templates. The thing that sets these templates apart from the others is the inclusion of Dynamic Component, which represents the width, depth, and height of the printable volume of a 3D printer. While this sounds like a great idea, the printers that are included in Dynamic Component are old printers that you are likely not using. Plus, the creation of a build volume should only be a three-second task and is something that we will cover in *Chapter 5, Modeling from Scratch Using Native Editing Tools* .

For the chapters where we go step-by-step through modeling together, I will start by mentioning the template that I use, when using SketchUp for Desktop.

Extensions

One of the big differentiators of SketchUp for Desktop is the ability to install **extensions**. Extensions are bits of code or small programs that run inside SketchUp that expand the capability of SketchUp beyond what it can do, out of the box.

For the purposes of modeling for 3D printing, this may mean adding extensions that allow us to more easily model organic geometry or that will automatically add width to single surfaces so that they are recognized by our slicing software. Or, it could mean installing an extension that will check our model to see whether it is solid and ready to print and, if not, tell us how we can fix it.

Here are a few extensions that you may want to look into as well as links to where you can download them. While there are hundreds of extensions out there, this list consists of a few that are relevant to 3D printing modeling workflows:

- **Solid Inspector**[2] from Thomthom (`https://extensions.sketchup.com/extension/ aad4e5d9-7115-4cac-9b75-750ed0902732/solid-inspector`): **Solid Inspector** checks your model to make sure it is solid and, if not, offers suggestions for repair or offers to fix the problems for you, automatically! The **Solid Inspector** tool is a part of the SketchUp Go version of SketchUp for Web, but in SketchUp for Desktop, it is available as a free extension.

- **SolidSolver** from TIG (`https://sketchucation.com/pluginstore?pln=TIG_ solidsolver`): This is an older extension, but it still works with newer versions of SketchUp. There are cases when **Solid Inspector** will not recognize a shape as solid or fix an error with a solid. In some cases, SolidSolver can help!

- **Joint Push Pull Interactive** from Fredo6 (`https://sketchucation.com/ pluginstore?pln=JointPushPull`): Joint Push Pull is a paid extension that is a huge time saver when you need it. While this extension offers a lot of functionality, the portion that directly connects to creating print-ready models is the "thickening" ability. This allows you to select a single surface and add thickness to it with a single click. This is great for creating solid geometry from a single surface.

- **Flowify** from Anders L (`https://extensions.sketchup.com/ extension/79bc42e4-d0d2-4a49-9b0a-1bd24ef3b3e7/flowify`): Flowify is an amazing extension that will allow you to distort one shape so that it follows the curvature of another. Imagine creating a solid model of a building, then telling it to wrap the building around a sphere. This is the sort of distortion you can create with Flowify.

At this point, I feel like I am only scratching the surface of extensions that can help with the 3D printing workflows that are out there. There are so many options to help create and check geometry, that it is hard to pick just a few. The one that is a must-have, however, is Solid Inspector[2]. We will actually be using it as a part of the step-by-step walkthroughs in the last few chapters of this book. For this reason (along with the fact that it is a free extension), I will assume that you have downloaded and installed Solid Inspector[2].

Up until this point, we have run through the different versions of SketchUp and what you should consider before you start modeling. There are a few additional items you may want to consider having around before you start modeling, however.

Real-world tools

While not a necessity, I find it nice to have a couple of measuring tools at my desk while I model for 3D printing. Unlike a lot of my other modeling projects, since these models will end up occupying space in the real world at some point, it can be nice to have a reference for them in real size. For this reason, I usually have these items at hand while modeling.

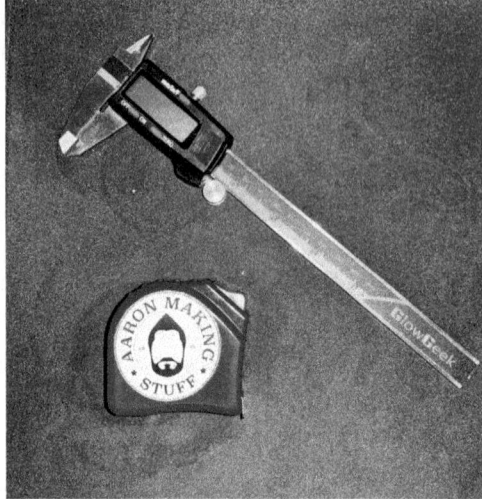

Figure 2.6 – My measuring tools for 3D print modeling

Tape measure/ruler

I tend to keep a small tape measure at my desk for reference when I model. I use it for non-3D print models, often using it to check the height of my desk or holding it out to see how wide a certain measurement is. For 3D printing, I tend to use it to check what size I want to print something. For example, the baby coffee cup that I modeled in *Chapter 1, Getting Started with 3D Printing and SketchUp*, was modeled at roughly full size. Before printing it, I scaled it down to a cup that was 1 ½" wide. I came to use this measurement based on holding my fingers in front of the tape measure and deciding how wide the little cup should be.

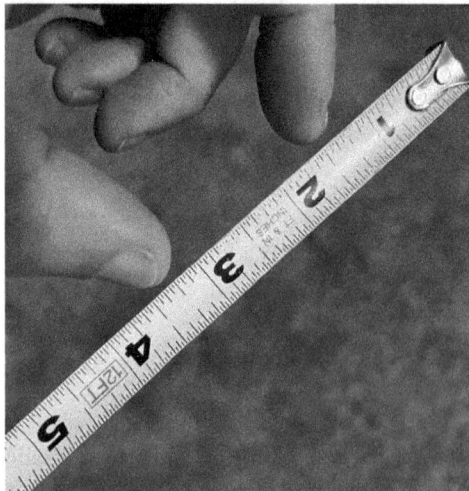

Figure 2.7 – Test measuring for output

Calipers

If you plan on modeling and printing a replacement part at any point (and everyone with a 3D printer does this sooner or later), then having a decent set of calipers is imperative. Yes, a lot of measurements can be done with a ruler. However, if you are going to print a piece that is going to interact with other items in the real world, then the accuracy that is afforded by using calipers is important. I regularly use them to check the width of a hole that a printed item will need to fit into or the thickness of the wall of a piece that I am replacing.

Other than these tools, you would want to keep any reference items around that may interact with your final print. By this, I mean things such as screws, or real-world parts that will connect or interact with the item that you are creating. Other than that, though, you should be ready to start modeling!

Summary

In this chapter, we went through SketchUp's capabilities and identified exactly what it will do, specific to the 3D print modeling process. This should be a great help if you have any confusion about what version you are currently using or if you are considering upgrading in the future.

After that, we looked at the three different versions of SketchUp that are available and walked through some of the steps you should go through to get ready to create print-ready models. This sets us up to start modeling together after we cover a few more basic concepts.

Up next, let's take a look at how to get files for printing in and out of SketchUp in *Chapter 3, Importing and Exporting .stl Files.*

3

Importing and Exporting .stl Files

You have probably seen that 3D printing involves passing files back and forth to go from an idea to a finished print. When it comes to generic, anyone-can-read-it files, `.stl` files are the file format of choice in the 3D printing world. This format is not, however, the default file format for SketchUp. SketchUp has its own file format (`.skp` files) and uses import and export functions to pass data in and out of `.stl` files.

Don't worry about having to simply read about this process, though. In this chapter, we will be getting our hands dirty (virtually) and using SketchUp to import and export a few files. Since this can be done in SketchUp for Web or SketchUp for Desktop, we will be taking a look at the steps involved in both!

In this chapter, we will cover these main topics:

- Understanding what makes up a SketchUp file
- Exporting files for 3D printing
- Importing files into SketchUp

Technical requirements

For this chapter, you will need your preferred version of SketchUp and the `3D Printing with SketchUp Chapter 3 - Baby Coffee.skp` SketchUp file from 3D Warehouse: `https://3dwarehouse.sketchup.com/model/a494bfeb-6a98-4772-8b1c-0903a73eb72d/3D-Printing-with-SketchUp-Chapter-3-Baby-Coffee`.

Understanding what makes up a SketchUp file

SketchUp saves its files in its own format as a `.skp` file. A `.skp` file contains much more than just the geometry we need for 3D printing. Before we dive into how to import and export geometry, let's spend just a few minutes looking at what makes up a SketchUp file, and where importing and exporting will fit into the default file format.

When you create a SketchUp file, regardless of the intended purpose of the final model, you create a file that contains a lot of information. When you think about all of the different designers that use SketchUp in their various industries, it is easy to imagine how many different models could be created. Architects, interior designers, furniture makers, and landscapers are all using the same software that you are, and they all create different models. This is not a problem, but it is something to recognize as we ask SketchUp to help us create geometry that we can send to our 3D printer.

While we may tend to think about the final geometry in the 3D printing workflow, it is important to realize that information about the location, materials, sections, and styles are saved in a SketchUp model as well. While we could spend the first half of this chapter going into everything that makes up a `.skp` file, let's look at the things that are important to the process of modeling print-ready models instead:

- **Geometry**: The edges and faces that make up a model
- **Groups and Components**: These are the containers that hold the geometry and define what is a printable solid
- **Units and Scale**: Basic information about the unit of measure in which the model is displayed, and the size of the geometry
- **Softening/Smoothing**: While geometry in SketchUp is made of rigid triangles, it is possible to view the geometry as smoothed out
- **Tags and Scenes**: These are useful for viewing your geometry as you model or return to work on your model over time

Again, this is only a small piece of what is saved into a SketchUp file, and the pieces that are directly connected to the modeling process, which we will be developing. Most of these pieces directly influence the information used to create and export geometry as well.

While it would be nice to be able to take the SketchUp file and directly import it into your slicer software, there is no slicer that I know of that will accept a `.skp` file. This is why SketchUp is equipped to export the file format that is most often used to transfer 3D models for 3D printing: the `.stl` file.

Unlike a .skp file, a .stl file contains only 3D geometry. All the other information about your 3D creation is discarded and you are left with just the basics – that is, the location of a bunch of interconnected triangles in 3D space. A .stl file does not care where this model is being built or what color the material will be, it is just simple geometry.

In the remaining chapters of this book, we will focus on creating a geometry that can be exported as a .stl file. For the remainder of this chapter, we will focus on the process of exporting the geometry you have created as a .stl file and see how to bring .stl files into SketchUp.

Exporting files for 3D printing

As you use SketchUp in your 3D printing workflow, you will be exporting a lot of .stl files. There will be times when your model goes from SketchUp to the slicer and the printer without issue. But sometimes, you will be jumping back and forth between SketchUp and your slicer, trying to get a model perfect for printing. Still, other times, you may not know that a model needs to be changed until after it has been printed! Regardless, you will need to know the process of exporting a .stl file. Fortunately, SketchUp makes this process as easy as possible, in any version.

Exporting a .stl file from SketchUp for Web

The process of exporting a .stl file from SketchUp for Web is the same if you are a Free or Go user. For this example, we need to open a SketchUp file, so we have some geometry to export. We will use the baby coffee mug from *Chapter 1, Getting Started with 3D Printing Using SketchUp*, as an example model, 3D Printing with SketchUp Chapter 3 - Baby Coffee.skp. If you have not downloaded it already, go ahead and do so now: https://3dwarehouse.sketchup.com/model/a494bfeb-6a98-4772-8b1c-0903a73eb72d/3D-Printing-with-SketchUp-Chapter-3-Baby-Coffee.

Let's start by exporting a .stl file from SketchUp by opening the downloaded file:

1. From the **Home** screen, click the **Open from Device** button.

2. Navigate to the downloaded file, select 3D Printing with SketchUp Chapter 3 - Baby Coffee.skp, and then click the **Open** button.

    The file will open, with the model of the mug filling the drawing area:

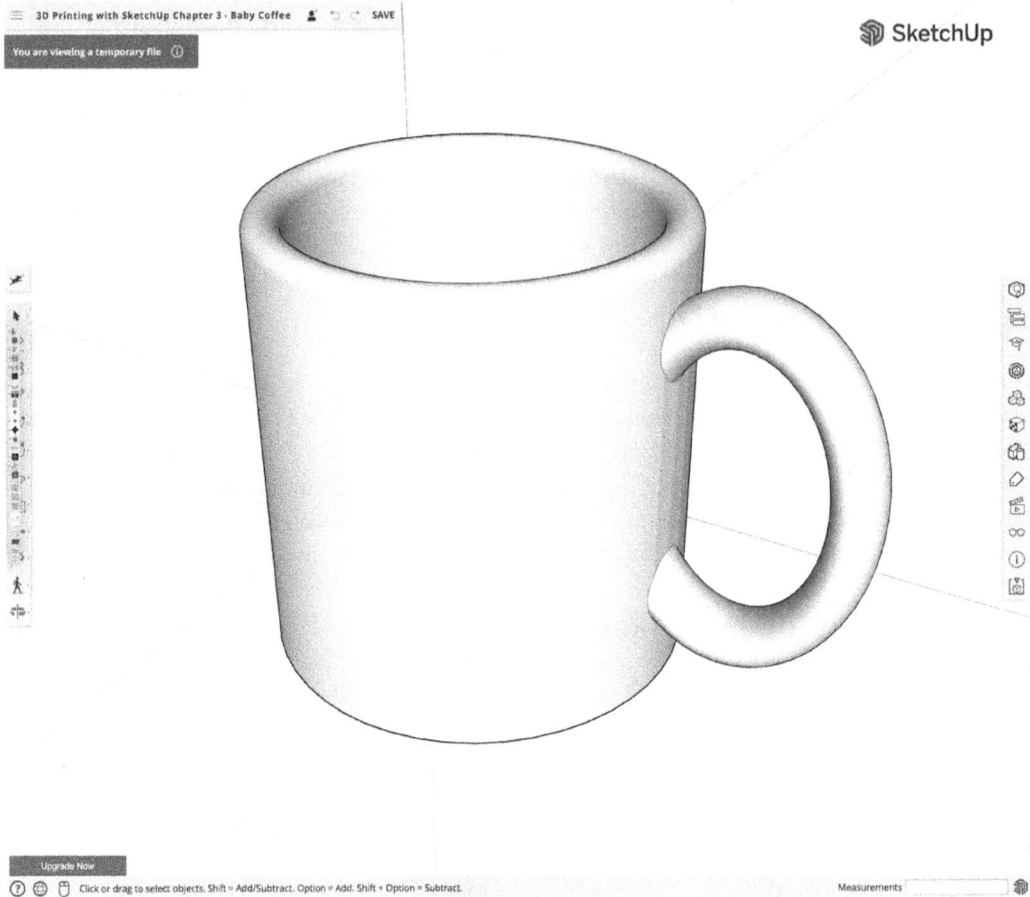

Figure 3.1 – Baby coffee mug open in SketchUp for Web

If we zoom out, however, we will see that there is a scale figure to the right of the mug and some loose geometry to the right.

3.   Zoom out using **Zoom Extents**:

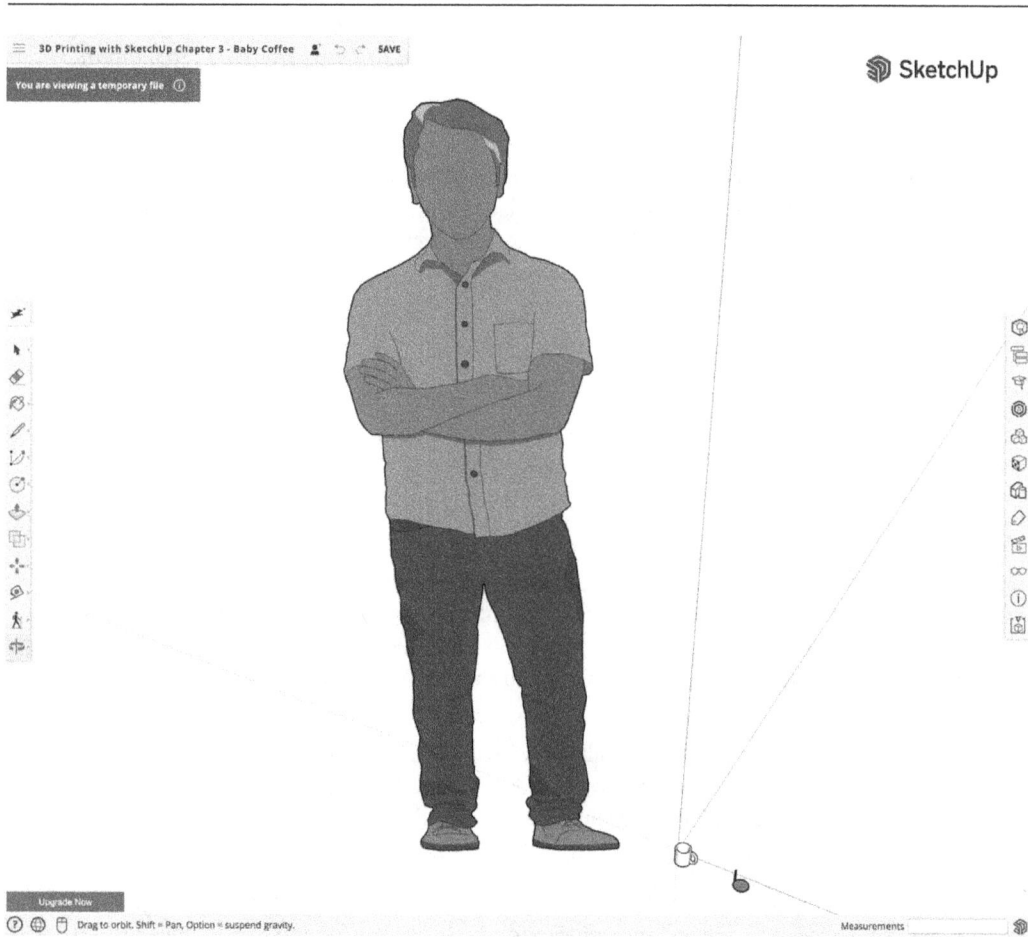

Figure 3.2 – Extra geometry in the model can be seen when we zoom out

This is an extremely important thing to notice. When we export from SketchUp for Web, a `.stl` file will be created from the entire model. If we were to have gone straight to the step where we export, our `.stl` file would have contained a whole bunch of unprintable geometry, along with our little mug. Let's get this model cleaned up before we export it.

4. Use the **Eraser** tool to get rid of the human component (also known as Niraj, the scale figure of SketchUp 2022) and the loose geometry on the right of the mug.

5. Click **Zoom Extents** again.

At this point, you should see only the mug. That is perfect. Now, we need to set the proper units for export. All the slicing programs that I use expect `.stl` files to import in millimeters. Before exporting, let's change the units of this model (which was created using inches).

6. Open the **Model Info** panel on the right and change **Length Units** from **1/2"** to **0.00mm**.

You can keep inches

If the program that will be importing this file is expecting the geometry to be reported in inches, then you can leave it set to 1/2". In general, the programs that I use in my 3D printing workflows all expect files to be in millimeter format (which makes sense since the units are smaller and more appropriate for the smaller geometry), so that is why this step was included in this workflow.

---

7. Click on the menu button at the top right, choose **Download**, and click **STL**.

    This will immediately start the download process for a file named `3D Printing with SketchUp Chapter 3 - Baby Coffee.stl`.

Automatic download

Depending on the settings of your browser, SketchUp for Web will prompt you for a download location and allow you to change the name of the file, or it may just download the file using the same name as the SketchUp file right into your default download location. Control for this behavior is completely up to your browser and should be changed to the method that you would prefer.

If you have your browser set to automatically download files, but you want this file to be named something different, you can choose to save the file with a different name using **Save As** before downloading or renaming the file after you have downloaded it.

That is all there is to downloading a `.stl` file from SketchUp for Web. The important things to remember are that your `.stl` file will include everything in the model and that it will export using the units displayed in the **Model Info** panel. Other than that, there are no export options to worry about at all! If you want more control over the file that you export, then you will have to look into SketchUp for Desktop, which we will do next.

Exporting a .stl file from SketchUp for Desktop

The process of exporting a `.stl` file from SketchUp for Desktop is pretty easy as well but gives you more freedom and control than SketchUp for Web. As we did in the *Exporting a .stl file from SketchUp for Web* section, we need to open a SketchUp file so that we have some geometry to export. We will use the same file, `3D Printing with SketchUp Chapter 3 - Baby Coffee`. If you have not downloaded it already, go ahead and do so now: `https://3dwarehouse.sketchup.com/model/a494bfeb-6a98-4772-8b1c-0903a73eb72d/3D-Printing-with-SketchUp-Chapter-3-Baby-Coffee`.

Let's start exporting a `.stl` file from SketchUp by opening the downloaded file:

1. From the **File** menu, click **Open**.

2. Navigate to the downloaded file, select `3D Printing with SketchUp Chapter 3 - Baby Coffee.skp`, and then click the **Open** button. The file will open, with the model of the mug filling the drawing area.

    If we zoom out, however, you will see that there is a scale figure to the right of the mug and some loose geometry to the right.

3. Zoom out using **Zoom Extents**.

    Unlike SketchUp for Web, SketchUp for Desktop will allow you to select the group or component that you want to export, so extra geometry does not matter. Often, when I model, I will have multiple versions of my model sitting side by side. If you want to clean it up and delete this extra geometry, you can, but you do not have to.

4. Use **Select** to highlight the coffee cup group.

5. Click on the **File** menu and choose **Download**. Then, click **Export**, and then **3D Model…**.

    This will open the export window. The first thing to look at is the **Format** field at the bottom of the window. We need to make sure that we are outputting a `.stl` file.

6. Click the **Format** dropdown and choose **StereoLithography File (*.stl)**.

7. Next, click the **Options…** button to the right of **Format**.

    This will display our export options. There are four options listed here. I will tell you my preferred setting based on exporting to the slicers that I use (*ChituBox* and *Cura*):

    - **Export current selection only**: I always have this box checked. This is what allows me to keep the other stuff lying around the model, as only the group I have selected is exported. Plus, with this option, I can export multiple `.stl` files from one model!

    - **File format**: I keep this set to **Binary**. The other option is **ASCII**. As far as I know, **Binary** is the most commonly used file format as it creates a more compact file. If you find that your slicer wants an ASCII file, change that here.

    - **Swap YZ coordinates**: This will change the *Y* and *Z* axes in your exported file, effectively rotating your geometry in the file. I always arrange my models in the slicer so that I am not overly concerned about what direction they face initially. I have never turned this on. If you find that your models import into your slicer and are always lying on their side, you can try checking this box.

    - **Units**: As I mentioned when we went through exporting using SketchUp for Web, the slicers that I use expect geometry to come in as millimeters. I keep this field set to **Millimeters**. You can change this to the unit of measure that your slicer is expecting.

> **Model units and export units**
>
> The nice thing about having a separate unit of measure set just for output means that you can model in whatever unit you like, and the file will still export in the units that your slicer needs. You can swap between units while you model (I often model in inches but change to millimeters when I prep a model for export) without any concern for the final unit because it is controlled here, and not by what you have modeled.

8. Once you have the output settings you need, click the **OK** button.

    The final thing we will look at is the filename. By default, the file being created will have the same name as the model, so right now, our export is going to be named `3D Printing with SketchUp Chapter 3 - Baby Coffee.stl`. In SketchUp for Desktop, we have the option to change this before we export, so let's take advantage of this and shorten the name.

9. Change the filename to `Baby Coffee.stl`.

10. Navigate to where you would like to save your new `.stl` file, then click the **Export** button.

You now have a `.stl` file of my coffee cup, ready to print! You may have noticed that the big difference between exporting a `.stl` file from SketchUp for Desktop and SketchUp for Web is that you have a lot more control. You can choose to export only a portion of your model and set the units for export as you create the file. If you are going to be doing a lot of printing, this feature alone makes SketchUp for Desktop a must-have!

Now that we have seen how to export a `.stl` file, let's take a look at the process of bringing one into SketchUp.

Importing files into SketchUp

Just like exporting, the process of importing a `.stl` file varies from one version of SketchUp to the next. While we will get into modifying and repairing imported files in *Chapter 7, Importing and Modifying Existing 3D Models*, right now, we will just cover the process of pulling `.stl` files into SketchUp. First, we will look at SketchUp for Web.

Importing .stl files with SketchUp for Web

The first thing to note is that the ability to import `.stl` files is an option that is only available with a paid subscription. This means that you have to be running SketchUp for Web for SketchUp Go or SketchUp Pro. The free version of SketchUp for Web does not include the ability to import `.stl` files. Knowing that, let's walk through a quick step-by-step guide using the `.stl` file you just created:

1. Start a new model in SketchUp for Web.

2. Click on the menu button at the top right, then click **Import**, and then **My device**.

At this point, you have the option to import a file from **Trimble Connect** as well. If you are using Trimble Connect as storage for files, you can import them directly (without the need to download them). Since we are importing the `.stl` file we downloaded from SketchUp for Web, we are importing from **My Device**.

This will bring up the **Import File** dialog, as seen in *Figure 3.3*:

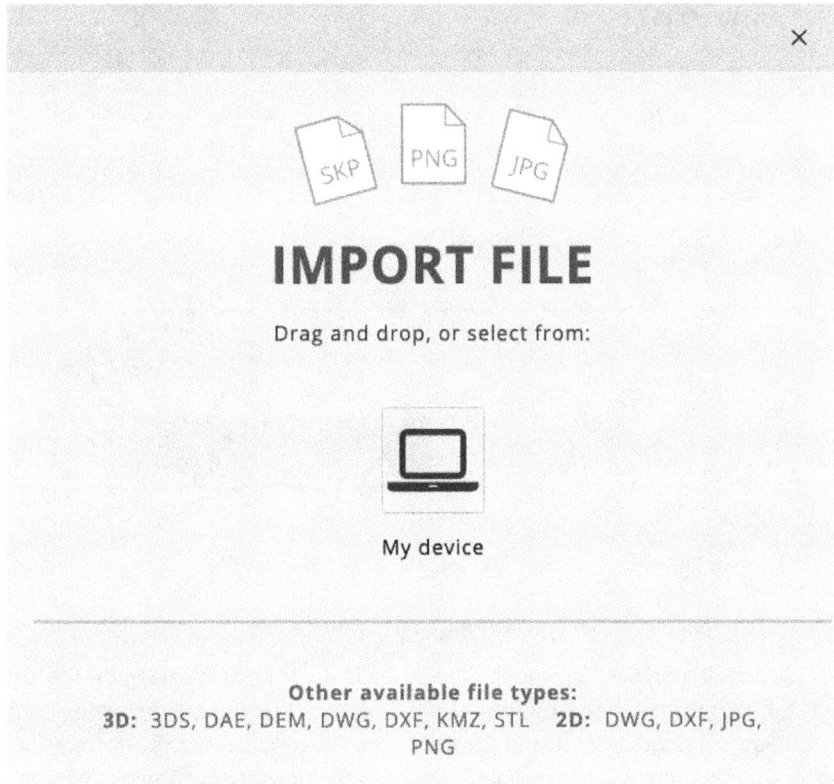

IMPORT FILE

Drag and drop, or select from:

My device

Other available file types:
3D: 3DS, DAE, DEM, DWG, DXF, KMZ, STL   **2D:** DWG, DXF, JPG, PNG

Figure 3.3 – SketchUp for Web's Import File dialog

As you can see, `.stl` files are listed as one of the types of 3D files that can be imported. You can import a file by dragging and dropping it onto this dialog, or by clicking the **My device** button and selecting it from your computer. Let's use the **My device** button to navigate to our downloaded file.

3.   Click on **My device**.

4.   Navigate to the folder where you downloaded the `.stl` file in the *Exporting a file for 3D printing* section and select 3D Printing with SketchUp Chapter 3 - Baby Coffee.stl.

5.   Click the **Open** button.

This will return you to the **Import File** dialog, with your selected file listed and ready for import.

6.   Click the **Import as Component** button.

When you start importing a `.stl` file, you will be prompted with some import options via the **Import STL** dialog:

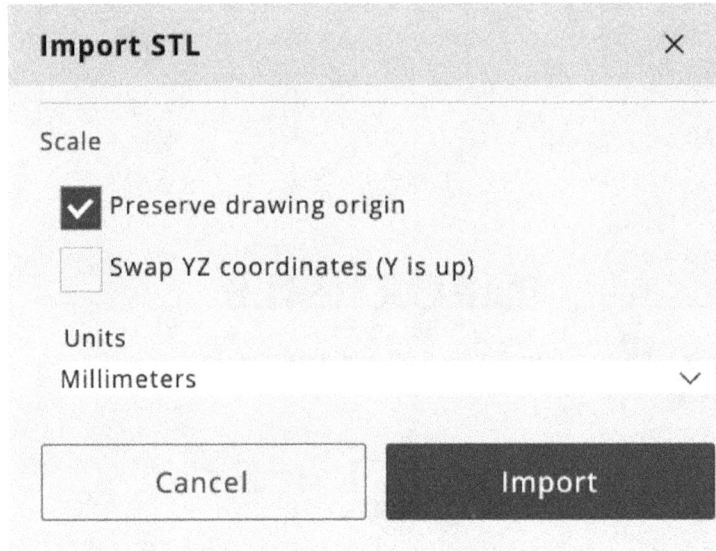

Figure 3.4 – Import STL options in SketchUp for Web

Before we finish the import, let's run through our import options:

- **Preserve drawing origin**: Keeping this checked will use the origin data in the file. Generally, this is OK to leave checked. You may need to uncheck it if you are importing a `.stl` file that you are not familiar with, and the origin point in the file is far away from the origin. Unchecking this box will assign a new origin to the imported geometry. We can leave it checked for our import.

- **Swap YZ coordinates (Y is up)**: Checking this box will swap the Y and Z axis information, effectively running your model on its side. There's no need to check this box.

- **Units**: This final field will allow you to set the unit of measure for the imported geometry. Since we know that we exported in millimeters, we should make sure that **Millimeters** is selected. If you are importing a model that was saved using a different unit of measure, be sure to choose accordingly.

7.   Click the **Import** button.

This will return you to the drawing screen with the `.stl` geometry attached to your cursor. The next time you click the left mouse button, the geometry will be placed into your model wherever your cursor is. Let's place our geometry near the origin.

8. Click on the origin in the model to place the imported geometry.

    This will import the file and display it as SketchUp geometry. You will notice that every face is broken into triangles, and nothing is smoothed out. This is normal for an imported .stl file (which saves all geometry as a series of triangles).

    You will also notice that, if you click on the model, the geometry is inside a component. Again, this is standard as far as how files will be represented, once imported.

At this point, the imported geometry is just simple SketchUp edges and faces and can be edited using native tools in the same way as the geometry created from scratch.

Next, we'll see how importing into SketchUp from Desktop differs from the process in SketchUp for Web.

Importing .stl files with SketchUp for Desktop

The process for importing a .stl file into SketchUp for Desktop is similar to what we just went through in SketchUp for Web but is different enough that it makes sense to step through it separately. Let's import the .stl file that we exported in the *Exporting a file for 3D printing* section of this chapter:

1. Start a new model in SketchUp for Desktop.

2. In the **File** menu, click **Import…**.

    This will bring up the import dialog. Before selecting the file, we'll look at the import options available to us.

3. Click the **Format** dropdown and choose **StereoLithography File (*.stl)**.

4. Next, click the **Configure…** button to the right of **Format**.

    This will bring up the **STL Import Option** dialog. The options in this dialog are very similar to what we saw in SketchUp for Web:

    *   **Merge coplanar faces**: Checking this box will make SketchUp for Desktop attempt to merge faces that are on the same plane. The .stl file will represent a single rectangle as two coplanar triangles. Checking this box before importing will merge those two triangles back into a single rectangle, resulting in less geometry after import. If you are planning on editing geometry imported from a .stl file, this is a good option to turn on. If you want to see all the geometry as it exists in the .stl file (which is how files are imported into SketchUp for Web), go ahead, and turn it off for this import.

    *   **Preserve drawing origin**: Keeping this checked will use the origin data in the file. Generally, this is OK to leave checked. You may need to uncheck it if you are importing a .stl file that you are not familiar with, and the origin point in the file is far away from the origin. Unchecking this box will assign a new origin to the imported geometry. We can leave it checked for our import.

- **Swap YZ coordinates (Y is up)**: Checking this box will swap the *Y* and *Z* axis information, effectively running your model on its side. There's no need to check this box.

- **Units**: This final field will allow you to set the unit of measure for the imported geometry. Since we know that we exported in millimeters, we should make sure that **Millimeters** is selected. If you are importing a model that was saved using a different unit of measure, be sure to choose accordingly.

5. Click the **OK** button.

6. Navigate to the folder that you exported to in the *Exporting a file for 3D printing* section and select `Baby Coffee.stl`, then click the **Import** button.

    This will create a new component with the imported geometry and place it near the origin (no need to place it in SketchUp for Desktop). Just like in SketchUp for Web, all edges will be visible, though you may see fewer if you chose to merge the coplanar faces.

That is everything we need to cover when it comes to the basics of importing. The process is simple, and you may be doing it regularly, depending on your 3D print modeling workflow. If you do think you will be doing this frequently and using SketchUp for Desktop, there are a few extensions that you may want to check out.

Extensions for importing files

As we just saw, you do not need any extensions to import a `.stl` file into SketchUp. If you want to take importing to the next level, however, these are a few extensions that you might want to investigate:

- **Skimp** (`skimp4sketchup.com`): This extension from mind.sight.studios adds a bunch of great controls for importing geometry. With Skimp, you can control how much detail you are importing by decreasing the detail, thus simplifying the geometry. You can also preview the model before committing to importing. This is a great tool, especially for larger, more complex files.

- **Transmutr** (`lindale.io/transmutr`): Transmutr from Lindale performs many of the same tasks as Skimp but adds a whole slew of additional functionality to the backend. If you need the option to simplify geometry and do any level of rendering or post-modeling work that is not for 3D printing, then you should take a look at Transmutr.

- **CleanUp**[3] (`https://extensions.sketchup.com/extension/046175e5-a87a-4254-9329-1accc37a5e21/clean-up`): CleanUp from Thomthom will not be used directly in the import process like the previous extensions but will make cleaning the geometry after importing quick and easy. With tools to merge faces and remove unneeded edges, CleanUp is a great extension to have installed for any workflow.

The process of importing a `.stl` file into SketchUp (for Web or Desktop) is fairly straightforward but is something you need to understand as it is something that you may end up needing to do regularly. Following the process outlined here should have you pulling files into SketchUp without issue!

Summary

In this chapter, we covered the basics of SketchUp's `.skp` and `.stl` files so that you understand the difference. You also saw the exact steps needed to export your model as a `.stl` file that can be used by your slicer. Finally, we walked through the process of importing `.stl` files so that they can be edited in SketchUp.

When it comes to using SketchUp as a part of the 3D printing workflow, exporting `.stl` files is an essential skill and something you will be doing regularly. Importing `.stl` files (or possibly other formats) may end up as a part of your workflow as well but won't be needed if you are generating your print geometry from scratch.

At this point, you are probably tired of talking theory and preparing to model. Well, you are in luck as the next chapter, *Chapter 4, Print-Ready Modeling and Scaling for Export*, will have us not only thinking about what is important when creating geometry for printing but doing some modeling together!

4

Print-Ready Modeling and Scaling for Export

I have used the term **print-ready** a few times already in this book without really defining what it means. When I say that a model is print-ready, I am saying that it is a manifold solid, and should be recognized as printable by slicing software.

Modeling print-ready geometry is not the default method of modeling in SketchUp. Due to the open nature of SketchUp, you can model whatever you want, however you want. Modeling print-ready geometry means being consistent in your process of maintaining solid geometry as you model. Yes, you could model in a more haphazard manner, and then spend time assuring things are solid afterward, but modeling with the intention of sending your geometry out as a solid will help you streamline your modeling process.

Fortunately, this process is not difficult to master. In fact, we should be able to cover the main concepts in this chapter, and still have time to tackle a lesson on scaling your models as well!

In this chapter, we will cover these main topics:

- 3D modeling groups and components
- Modeling and scale
- Considering wall thickness and support

Technical requirements

This chapter will have some hands-on work in SketchUp. These steps can be performed in any version of SketchUp, so you will need access to SketchUp along with the 3D Printing with SketchUp Chapter 4 - Scaling.skp file from 3D Warehouse: https://3dwarehouse.sketchup.com/model/9cd6e76f-cb4a-454b-acdc-1c5f1cadb9ce/3D-Printing-with-SketchUp-Chapter-4-Scaling.

3D modeling groups and components

As you are probably already aware, in order for SketchUp to consider a collection of geometry as a solid, the geometry must be held in a group or a component. Let's do a quick exercise to dive into exactly what makes an acceptable solid:

1. Start a new file in SketchUp.

2. Use the **Rectangle** tool to draw a rectangle, then use the **Push/Pull** tool to pull the rectangle up into a box.

3. Triple-click the box to select everything.

4. If it is not already visible, open the **Entity Info** panel.

    At the top of the **Entity Info** panel, it will tell you exactly what is selected. Currently, you have **18 Entities** selected (6 faces and 12 edges). What you have selected is not a solid and is not printable. Let's fix that.

5. With all entities still highlighted, right-click on the geometry and click **Make Group**.

    Notice that the **Entity Info** panel now reports that you have **Solid Group** selected. This group, if exported as an `.stl` file, could go to a slicer and then to a 3D printer. This group is a print-ready model.

> ### Groups or components
>
> In this example and many of the examples in this book, I use groups to create solids. You could use components for this process, just as well. Unlike groups, components require you to specify a name each time you create one, and every copy of a component is connected – modifying one instance will update all copies. Unless there is a reason to maintain a connection between copies, I will just use a group, instead.

Just throwing geometry into a group will not make it a solid, however. In fact, it is terribly easy to modify a solid so that it is no longer recognized as a solid at all. Let's intentionally cause some problems with this group, so that you will know what to look for if your models ever report back as not being solids. To do this, we are going to recreate the top three issues that cause groups to not be solid.

Holes in geometry

Possibly the most common issue with groups not being solids is that there are holes in the geometry. Obviously, when you think about geometry being a solid, a hole in the outside is a problem. With our box example, let's remove a side and see what happens:

1. Double-click on the group.

2. Select the front face of the box and press *Delete*.

3. Click outside of the group to close it, then select it again.

Notice that the **Entity Info** panel now reports it as **Group** (rather than a solid group). This is because the faces do not connect to form a completely closed solid.

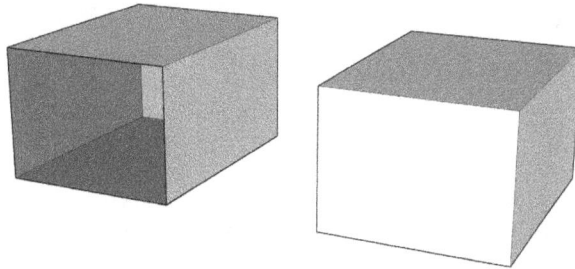

Figure 4.1 – A group with a hole on the left and a solid group on the right

To fix this issue, you need to track down the missing faces and fill them in. In some cases, it may be something as simple as a missing face, like in our model, or it may be two faces that almost meet but slightly miss each other, forming a small hole between their edges. In the case of our example, it is as simple as entering the group and drawing a new edge along the hole (or tapping **Undo** a couple of times to put the deleted face back).

Interior geometry

Another rule of a solid is that all the geometry is on the outside. This means that every face inside the group must form some portion of the exterior of the model. No faces (or edges, for that matter) should exist inside the solid. To see an example of this, let's duplicate the front face of our box:

1. Double-click on the group.
2. Use **Push/Pull** plus the modifier key to create a new face (*Option* on Mac and *Ctrl* on Windows) to pull a new face on the front of the box.
3. Click outside of the group to close it, then select it again.

    Notice that the **Entity Info** panel again reports it as **Group** (rather than a solid group). This is because of the interior face in the group.

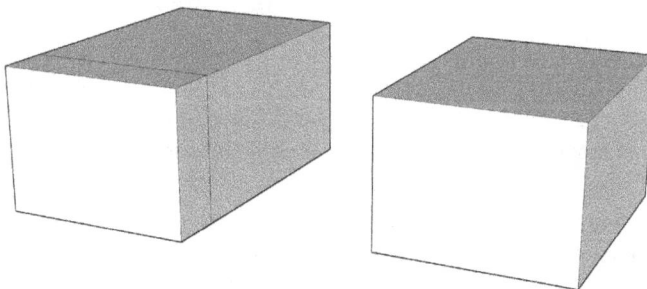

Figure 4.2 – A group with interior geometry on the left and a solid group on the right

As a rule of thumb, any face in the model should only connect to one other face at any given edge. If this happens, then there is no way that any geometry can be left inside the solid.

The solution to fix this issue can be a little trickier to execute but will consist of finding the extra geometry and removing it from the model. In the case of our example, however, it is as simple as clicking on **Undo** a couple of times.

Stray geometry

Another issue that can happen that will prevent your groups or components from being recognized as solid is loose geometry. This can be a face or an edge lying around outside of the geometry. It's pretty much the same as excess interior geometry, but on the outside. For me, this is often an extra edge that I was using as a reference or trimmed off and forgot to erase. Let's add one line and see how our group is no longer solid:

1. Double-click on the group.

2. Use **Line** to draw an edge anywhere. This edge can be connected to the geometry, as shown in *Figure 4.3*, or can be completely disconnected.

3. Click outside of the group to close it, then select it again.

Notice that the **Entity Info** panel again reports it as a **Group** (rather than a solid group). That is because there is now geometry in the group that is not a part of creating that single, manifold solid. Every edge in a solid group is connected to only two faces. We added an edge that is connected to no faces, so the group is no longer solid.

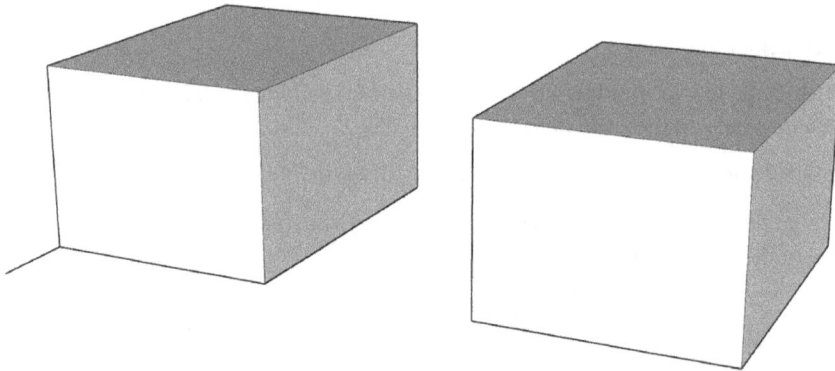

Figure 4.3 – A group with stray geometry on the left and a solid group on the right

Fixing this sort of problem is as simple as getting rid of the extra geometry. The tricky part is finding it. In this example, you can enter the group and use **Eraser** to eliminate the superfluous line or tap **Undo** a few times.

> **Hiding geometry won't help**
>
> Some may think that the solution to extra geometry could be to hide it, but that will not help. When deciding whether a group is a solid or not, SketchUp looks at all geometry, regardless of tag, visibility, color, or even connection to the rest of the geometry in the group. If the geometry is anywhere in the group, seen or unseen, it is considered to be a part of the solid.

Depending on the version of SketchUp you are using, you may end up using **Solid Inspector** to find and fix issues like this for you. This is great, of course, but it is a good idea to know the reasons why a model may not be solid in the first place, so you can avoid the issues altogether while modeling.

Now that you understand how to keep your geometry in a solid group, let's talk a little bit about how scale plays into creating a printable model in SketchUp.

Modeling and scale

In general, I recommend that everyone model at full scale in SketchUp. The fact is, SketchUp does not care what size you model something. The only thing that matters is what size your model is when you export the `.stl` file.

In some cases, this might not make much of a difference. If you are modeling a replacement knob for your stove or a pencil cup for your desk, you will likely just model at full size, then export the model, as is. If you plan to model something large, like your house or a spaceship, then you will want to model full size, then scale the model down before exporting.

Let's hop into the example model for this chapter and see how scaling works (if you don't have `3D Printing with SketchUp Chapter 4 - Scaling.skp`, now would be a great time to download it from the 3D warehouse: `https://3dwarehouse.sketchup.com/model/9cd6e76f-cb4a-454b-acdc-1c5f1cadb9ce/3D-Printing-with-SketchUp-Chapter-4-Scaling`):

1. Open `3D Printing with SketchUp Chapter 4 - Scaling.skp` and click on the scene, **Scale House**.

    We will start simple with this example and scale it down so that it will be printed. This model is a full-size house that will never fit onto the build plate of any printer. At this point, we could use the **Scale** tool and change the size of the whole model to 1%, or we could export the building full size and use our slicer to make it fit on the build plate. Alternatively, we can take control of the scaling process and tell SketchUp exactly what size we want this house to be when we export it.

2. Use **Select** to double-click and enter the house group.

3. Orbit around the model so that you can see the underside of the house.

    We will use the tape measure tool to scale the house down so that it is exactly 2" wide.

4. Choose the **Tape Measure** tool and toggle **Create Guides** off (*Ctrl* in Windows or *Option* in macOS).

5. Click on the back-right corner of the house and then on the back-left corner.

6. Type 2 " and press *Enter*.

    At this point, SketchUp will confirm that you want to scale the group down.

7. Click the **OK** button.

    At this point, the model of the house has been scaled and is ready for output. If you want to look at the new, scaled house, you may have to zoom in a bit, since it is less than 1% the size of the original house.

That is all there is to scaling a group so that you can export and print at a smaller size! This does, however, bring up a few issues with scaling and detail. Let's talk a little bit more about scale and detail with the next scene in this model.

Click on the scene's **Scale Detail**. This will bring up two scaled house models.

Figure 4.4 – On the left are real-world details, on the right are scaled up details

In *Figure 4.4*, the one on the left is a copy of the model we just saw, and on the right is the same model but with some of the details enlarged.

The details on the house on the left may seem perfectly acceptable until you see just how small they are. If we look at a detail, like the trim around the garage door, it looks correct in the model on the left, but if you check the width of that trim, you will find that it is just over 0.3 mm wide and only protrudes from the face of the wall a little over 0.05 mm. Considering that an FDM printer may print this with layers that are 0.2 mm, that makes the trim detail less than a layer thickness tall. That is super small.

If you look at the same detail in the model on the right, you will see that the trim is around 0.7 mm wide and over 0.2 mm off the face. Yes, it looks a little clunky and simple in the model, but take a look at how this looks when it is printed in *Figure 4.5*:

Figure 4.5 – The houses from Figure 4.4 printed on an FDM printer with a 0.2mm layer height

Notice how much detail you can see in the house on the right! The garage trim that we looked at before is invisible in the house on the left, while it is clearly visible on the right. Also, the columns in the front of the house were so thin in the house on the left that they broke while trying to remove the support material.

All of this is not to say that you cannot print fine detail. I should also acknowledge that printing this model in SLA would maintain some of the finer detail that is lost in an FDM print. The important thing to take away here is the level of detail you are putting in your model, and whether it is right for the final output (geometry going to a 3D printer). At this scale, there is no reason to add the doorknob to the front door, for example, since it will be too small and end up eliminated by the slicer.

Something else to consider as you print is the size of the actual model. Details will naturally be more visible in a print when the print is larger. Scaling models, of course, come with a whole new set of potential issues. Let's move on to the next section and talk about how scale affects your modeling workflow.

Scaling while modeling

One function you can take advantage of when modeling in SketchUp is the fact that component instances are connected. This means that anything you change in one component will change in all its copies, including scaled copies. If you were to copy a full-size component and scale the copy down to the size you want to use for output, you could jump back and forth, making changes to either component, effectively making changes to the full-size version or the scaled version at the same time!

For this example, we will take a look at a model I created and printed at two drastically different scales. I printed the rocket model in *Figure 4.6* twice. I printed it once on an FDM printer in three parts that stack up to a 13" tall print, and again on an SLA printer at 1 inch tall. Both prints came out perfect, and both came from the same model, as shown here:

Figure 4.6 – 13" print on the left and 1" print on the right from the same SketchUp model

Let's take a look at how to create two different-size prints from one model:

1. Click on the **Rocket** scene in 3D Printing with SketchUp Chapter 4 – Scaling.skp.

    This rocket is a 13" tall version of the rocket. To get a 1" tall version, we could rely on the slicer to scale it, or we can make the group into a component, and copy and scale the copy. Doing so will allow us to make changes to either instance of the component, and both will be updated.

2. Right-click on the group and choose **Make Component...** from the context menu.

3. Name this new component Rocket, then click the **OK** button.

4. Make a copy of the rocket using the **Move** command.

5. Select the new copy and start up the **Scale** tool.

6. Select the grip at the top center of the rocket, tap the *Shift* key to scale uniformly, then type 1" and press the *Enter* key.

   This has created two copies of the rocket that are exactly 13" and 1" tall. At this point, they could be exported as .stl files and sent to your slicer. Before that, however, let's say that we want to make a change to the geometry. I think that the top cone should be a little taller and pointier. Let's stretch that out a bit.

7. Switch to the **Select** tool and double-click to enter the taller rocket.

8. Click the **Move** tool, then hover over the very top point of the cone. When the endpoint highlights in green, click on it.

9. Tap the up arrow key and move your cursor up the screen.

   As you stretch the cone taller, keep an eye on the smaller rocket. Have you noticed that the cone in the smaller copy is changing as well? This is a huge help for modeling in two sizes at once.

10. When you like the shape of the nose cones, click to finish the **Move** command, and then use **Select** to click outside of the component to close it.

There may be situations where you want to maintain two different scales of a model. The most common one that I have come across is keeping one full-size model while at the same time maintaining a version that I know will fit into my printer. With two copies of the component, I can model in real-world units in the larger copy while keeping an eye on the scaled version to make sure that it is still printable.

Additionally, this will help you steer clear of the dreaded "small face" limitation that modelers can run into in SketchUp. SketchUp was originally created to allow people to model building-size models and struggles to create very small faces (such as faces that are less than 0.0001" across). Faces that small can exist in a SketchUp model, it's just that SketchUp has a hard time creating faces when they get that small. By working in a larger size copy of the geometry, you prevent yourself from running into the issue, while still creating a model with tiny faces that you need for the .stl export.

> **Will your model fit in your printer?**
> Something you can do to verify that your model is print-ready is to see that it will fit in the volume of your printer. One way to do this is to model a box the exact size of your print volume and make it into a component. If your model fits inside the box with nothing sticking out the sides, then you know your printer can print it. If you like this idea, you can even save it as a part of a template, so that you can always check your models.

Scale is an important piece of print-ready modeling. It can be tempting to leave scaling off until later, or let the slicer take care of it, but with a little bit of forethought and planning, modeling the scale can be just as easy as modeling full-size and can prevent you from having to resize your model downstream.

Up next, let's look at a few more factors to consider when attempting to create print-ready models, specifically, wall thickness and supported geometry.

Considering wall thickness and support

I do have to admit that slicing software has come a long way in the past few years. I remember when I started 3D printing, the software I was using was barely able to do much more than generate support for overhanging geometry. Nowadays, it can repair meshes, automatically orient models, and, in some cases, it can even print non-manifold geometry.

While I am all for taking advantage of this software when it makes sense, there are some things that you may want to control yourself, as you create your model, such as creating your model with proper wall thickness and creating geometry that will reduce the need for support.

Modeling wall thickness

Let me start by describing exactly what I am talking about here. I am using the word **wall** in reference to any geometry that creates the outside geometry of your model. To be properly manifold and printable, walls must have some thickness. They cannot exist as simple faces. Take the example in *Figure 4.7* as an example of what is print-ready and what is not.

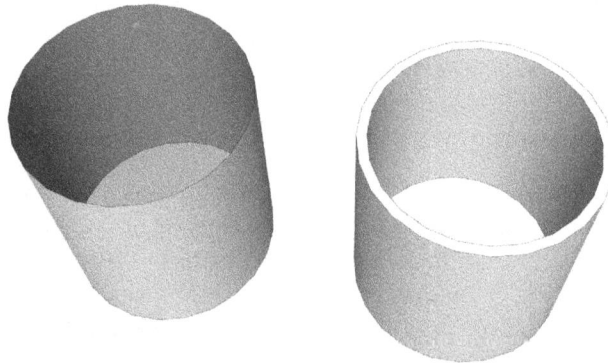

Figure 4.7 – On the left is a cup shape with walls with no thickness (un-printable),
and on the right is a cup with walls with thickness (printable)

As you model geometry, just remember that any face you create should be the outside shell of a closed-in geometry, so things such as single faces will not be printable.

So, if you want to add thickness to something such as this cup, what is the best dimension to use? If you are printing on an FDM printer, here is some simple math that will allow you to figure out the minimum wall thickness to use. If you take the width of a single line, then multiply it by two times your shell layers, you will get the minimum thickness of a wall (wall width = line size x 2 x shell layers). If I am printing with a 0.4 mm head (thus 0.4 mm lines) and I am printing two layers of shell, I would want a wall that is at least 1.6 mm thick (0.4 x 2 x 2).

If you are printing on an SLA printer, the answer is not quite so simple. I have read that you should be able to print a wall as thin as four times a single layer height. Personally, I have never messed with printing anything thinner than 1 mm. While I believe that I could get thinner than this, I worry about the durability of that geometry and removing support material from something that thin. Up to now, a minimum wall thickness of 1 mm has worked well for me.

Considering supports

I remember a time, way back, when I added support structures to my 3D model so that I could assure that overhanging geometry was properly supported. Back then, slicers were not very dependable and had few options when it came to setting up supports. As I started writing this chapter, I tried hard to think of a geometry I might print where I would want to manually add support geometry, and I could not come up with a single example! Today, I believe that I am better off editing support settings in my slicer to get perfect support rather than messing with creating extra geometry in my model. This does not mean, however, that you should not keep support in mind when you are creating your model.

One of the big things to think about is what support your geometry will require and whether there are simple changes you can make to limit the amount of support you will need to print it. While there are some things you will want in your model that will require support, you may be able to limit or eliminate support by modeling intentionally.

Consider printing dice. If you were to print a dice (or any cube or box shape) lying flat on the print bed of an FMD printer, you may be able to print without any support at all. If you wanted to add rounded or chamfered corners to the dice, however, you may end up needing some support. Consider printing the four cubes shown in *Figure 4.8*:

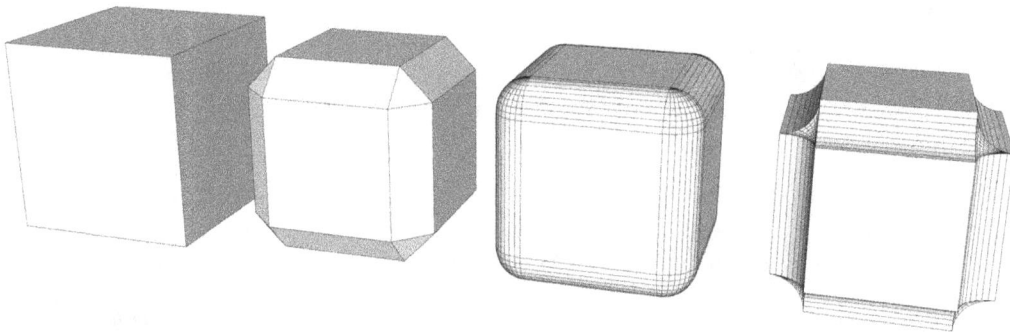

Figure 4.8 – Four cube shapes with all edges visible (no smoothing)

The shape on the far left would probably print just fine as all the walls rise directly from the print bed. The second, with 45-degree chamfers, would be OK, assuming your slicer was set to support geometry that angled less than 45 degrees. The third box is most likely to print well because the convex corners will allow your printer to build up without the need for support. If you were to try to print the fourth dice, however, you would probably need support material, as the concave corners create an extreme overhang above the print bed. To see this clearly, let's look at the cross sections of these same four shapes in *Figure 4.9*:

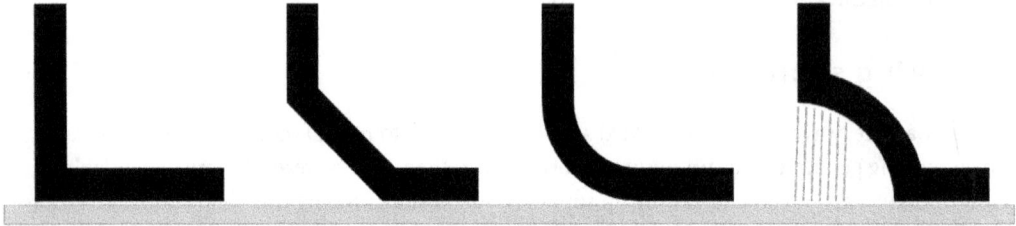

Figure 4.9 – Square, 45-degree, convex, and concave corner cross-sections on a gray build plate

Assuming your slicer is set to support geometry that angles more than 45 degrees, the only geometry that would need to be supported is the concave corner on the far right of the preceding figure. As you design geometry in SketchUp, you should keep this in mind and stick to corners that are 45 degrees or convex if they're not a perfect 90-degree angle.

When it comes to modeling the perfect print-ready geometry in SketchUp, there are quite a few factors to keep in mind. If you start modeling by making sure that you are creating manifold solids with some thickness to the walls and limiting unsupported overhangs, you will create models that are ready to print and decrease the amount of work that needs to be done in the slicer!

Summary

Creating models that are ready to go from SketchUp to the printer (with only the briefest of stays in the slicer) is not difficult and is something that you should strive for as you develop your 3D printing workflow.

In this chapter, we saw what it takes to create a solid group and how to use SketchUp to establish real-world and printed scales. Additionally, we saw some examples of considerations, such as wall width and unsupported geometry, that can be addressed with a little bit of forethought and planning.

Up next, let's model some print-ready models together in *Chapter 5, Modeling from Scratch Using Native Editing Tools*.

Part 2: Modeling for 3D Printing

The second part of the book is all about getting hands-on with SketchUp. This part will show you how to create, edit, import, and export models specifically for 3D printing.

In this part, the following chapters are included:

5

Modeling from Scratch Using Native Editing Tools

One of the things that initially drew me to SketchUp was how much I could get done with the native set of simple modeling tools. While I eventually mastered several advanced modeling workflows and the use of many extensions, the basic, simple modeling tools are still key to SketchUp modeling.

In the Free version of SketchUp for Web, I am taken back to those early days of SketchUp use, with access only to basic modeling tools. In this chapter, we will use only the native modeling tools in SketchUp for Web to create a print-ready model from scratch.

> SketchUp version
> While the exact steps in this chapter will be from the Free version of SketchUp for Web, you should be able to follow along in either the Shop version or SketchUp for Desktop.

In this chapter, we will cover these main topics:

- Creating basic geometry
- Editing geometry using native tools
- Preparing your model for 3D printing

Technical requirements

In this chapter, we will be modeling in the Free version of SketchUp for Web, but you can follow along with any version of SketchUp. You will also need the sample file for this chapter, `3D Printing with SketchUp Chapter 5 - Solids.skp`, which is available at `https://3dwarehouse.sketchup.com/model/e8362510-9421-49a1-bbe3-fa19405c0373/3D-Printing-with-SketchUp-Chapter-5-Solids`.

Creating basic geometry

For this example, we will model a print-ready model from scratch using only the tools available in the Free version of SketchUp for Web. This is modeling in its purest form. With this setup, we will have access only to the basic native toolset. If you happen to have a Go or Pro subscription, you can follow along in the full version of SketchUp for Web, as the UI and the commands are the same. If you prefer to follow along in SketchUp for Desktop, that should be easy enough, but be aware that any UI shown in the images of this chapter may look different from yours.

Now, all we need is an idea of what to model. Looking around my desk, the glass jar that I use as a pencil holder caught my eye. It is way too big for the few pencils and pens that I have on my desk, and to be honest, I have no idea where it came from. It seems like the perfect opportunity to have something that I designed and created on my desk instead.

Design time

I have never designed or printed a pencil cup before, and now that I think about it, I really have no idea how big they are! There are cases where measuring something in the real world will mean measuring with a tape measure or laser. Fortunately, this is a pretty simple real-world item to measure, as it will end up just sitting on my desk! Rather than recreating the dimensions of the glass jar I have been using, I will purposely make a new cup, specifically for the few items that need to go into it. My first step was to lay down the small collection of markers, pens, pencils, and craft knife that I have at hand and see how wide a cup would need to be.

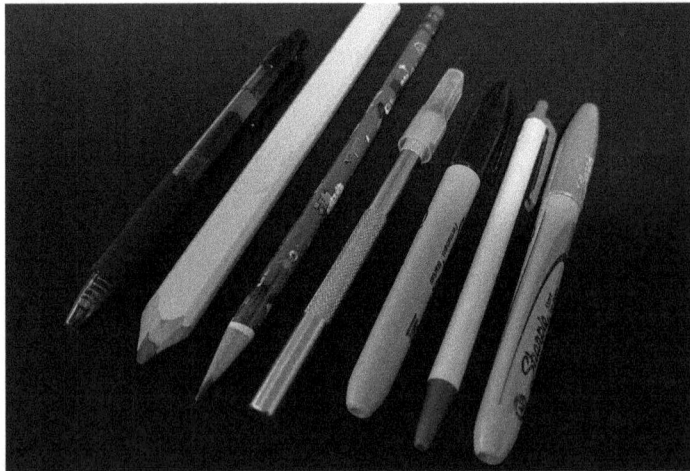

Figure 5.1 – The items that will need to fit in the new pencil cup

Now, I need to figure out how wide a cup would need to be to hold all these items. If I were to grab these items in my hand, I could get an idea of how wide I need this cup to be. Keeping in mind that

I want the items to fit loosely so that there is plenty of space to insert and remove them, as well as to allow for additional items to be added, I end up deciding that 3" should work pretty well.

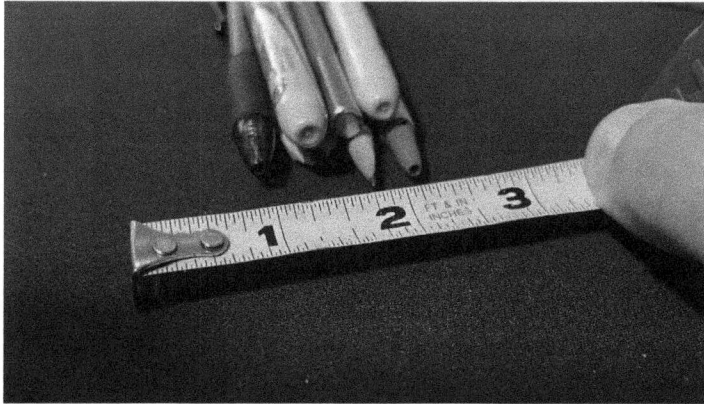

Figure 5.2 – Measuring the width of the new pencil cup

As for height, I want the cup to be tall enough that the items are held upright… something past half the height of an average drawing tool should work. Based on another quick measure, I come up with a height of 5".

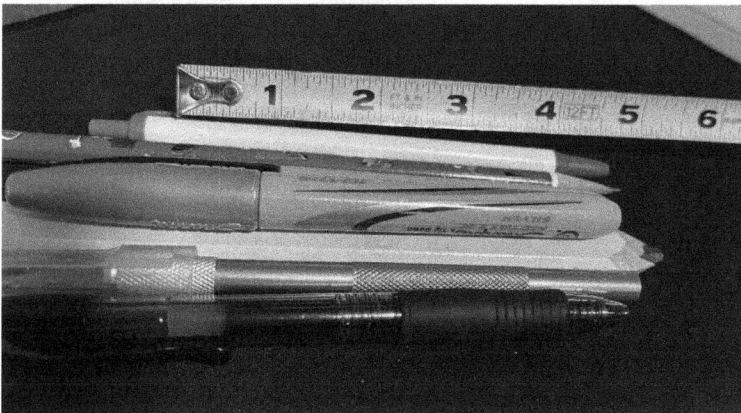

Figure 5.3 – Checking the potential height of the new pencil cup

The last measurement I need before I start modeling is the wall thickness. I could guess or hold up my calipers to get an idea of how thick this needs to be, or I can measure a real-world item. While I shied away from measuring my existing glass jar for most of the measurements, I think that the thickness of the lip might give me a good idea of how thick the wall of my new pencil cup should be. A quick measure with my calipers lets me know that somewhere around 4 mm should work.

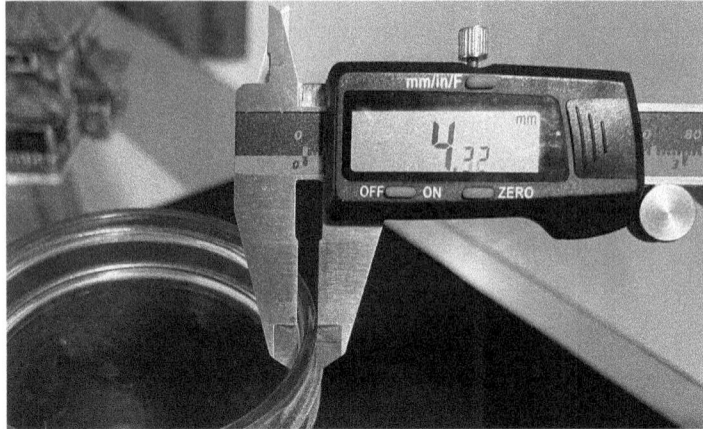

Figure 5.4 – Measuring the thickness of the glass jar

Now that I have the measurements, I just need to decide what I want this thing to look like. Yes, I could just model a round cup shape and call it good, but part of the fun of designing something from scratch is the ability to make it something unique!

I happen to have a spool of yellow-orange filament left over from a sample pack I ordered when I first got my FDM printer. While my red, blue, and green filaments are pretty much gone, this is left over because I don't love the color. It looks a bit like cheese. And that is what I will do with it! I will design my pencil cup as a large block of cheese!

I did a quick sketch of what I want my final pencil cup to look like.

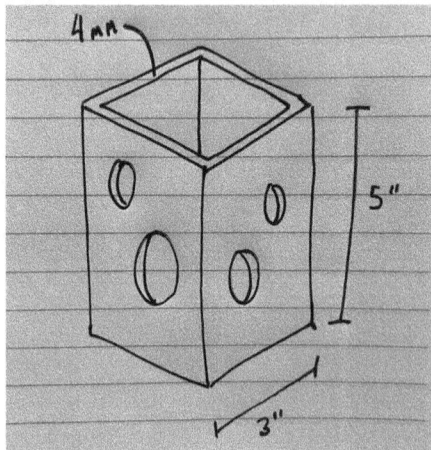

Figure 5.5 – A quick sketch of my soon-to-exist pencil cup

> **Straight to modeling**
>
> It is not a requirement to hand-draw your idea up before jumping into SketchUp to model. Since I cannot show you the image I have in my head, I rely on a quick sketch to share my vision before we start modeling.

Now that we have a plan, we can start modeling! Let's get into a brand-new model in SketchUp for Web and start modeling by following these steps:

1. Open SketchUp for Web in a new tab in your browser.

2. Click the **Create New** button.

    Let's build this model from the ground up. We will start by creating the 3" x 3" square on the ground and then work our way up.

> **Picking your units**
>
> Feel free to set up SketchUp for Web to use your preferred units at this point. As we saw in *Chapter 2, Setting Up Your SketchUp Environment for 3D Printing*, we will be changing units once we are ready to output, but while we are modeling, we can have it set to the unit that we prefer. Plus, SketchUp will accept any unit as we model if we define the unit of measure as we input. This means that we can have our units set to inches but define a dimension of 4 mm at any time by typing 4mm.

3. Click the **Rectangle** tool, and then click on the ground, near the origin.

4. Drag the cursor up and to the right, then type 3 " , 3 ", and press *Enter*.

    This should give you a perfect square, like the one in *Figure 5.6*:

Figure 5.6 – A square on the ground plane

As this is the base of our pencil cup, we might want to give it a little more thickness than the rest of the walls of our cup. Let's make the base twice the thickness of our walls.

5.   Click **Push/Pull**, and then click on the rectangle.

6.   Start moving your mouse up the screen, type 8mm, and then press *Enter*.

Now, our square has some depth! You should be seeing geometry just like what is shown in *Figure 5.7*:

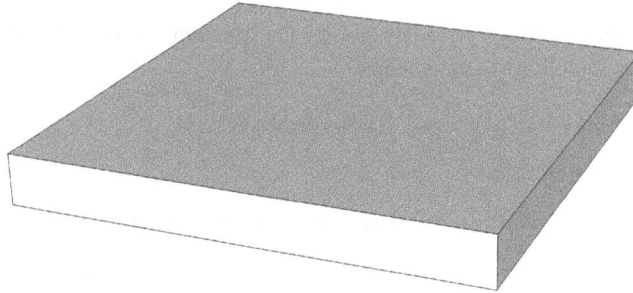

Figure 5.7 – The base of the pencil cup

The next step is to create the thickness of the walls. This is a simple offset of the square at the top of the base we just created.

7.   Click **Select**, and then click the top face of the base.

8.   Click **Offset**, then click the edge of the selection, and move your cursor toward the center of the square.

9.   Type 4mm, and then press *Enter*.

This will give us a rim around the square, as seen in *Figure 5.8*:

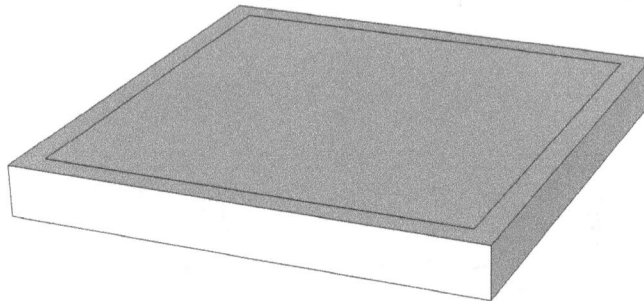

Figure 5.8 – The base with the wall width offset

The final step in creating the initial geometry is to pull the walls up to their full height.

10.   Click **Push/Pull**, and then click on the offset geometry we just created.

11.   Move your cursor up the screen, type 5 ", then press *Enter*.

As you perform this step, you may realize that we just created a pencil cup that is 5" and 8 mm. If the goal is to match the initial drawing, then we need to pull the walls up 5" minus 8 mm. This could be done by doing some conversion and math, but we will keep it simple and do a second push/pull, and then lower the top of the cup back down by 8 mm.

12. Click on the top lip and move your cursor down the screen. Then, type 8mm and press *Enter*.

With that, we have completed the initial geometry of our pencil cup.

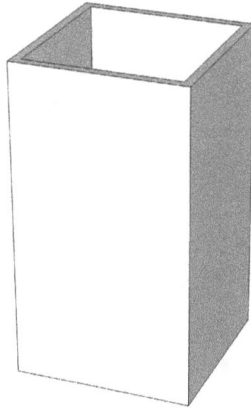

Figure 5.9 – The initial pencil cup geometry created!

At this point, you should probably click the **Save** button at the top of the screen and save your progress. We have more to do, but it is wise to save regularly. I called my model Cheese Cup, but you can save your model under whatever name you prefer.

At this point, we have created basic geometry for our model using the native commands in the Free version of SketchUp for Web. Next, let's add a few details (specifically the holes) using more native commands!

Editing geometry using native tools

The next step in creating our cheese-inspired pencil cup is adding random holes to the sides of the cup. This should be pretty simple, using some of the same tools we used to create the initial geometry.

Let's start by drawing a few circles onto one of the sides of the cup. While we want the location of the circles to appear organic and random, we need to be intentional about their placement. Since we want our cup to be functional, I will keep the holes up, away from the bottom of the cup, so that my pencils don't fall out the bottom. Additionally, I want to make sure that the holes are not too close to the edges. If I have a circle right up against the edge, it will interfere with the wall geometry of the next wall as I push the circle through.

After we create the circles, we will push them through to the opposite side of the cup. The easiest way to make sure that my circles will not lap over the side wall is to draw them on the inside face. This way, if a circle goes all the way to the edge, I know that it is still 4 mm away from the outside of the adjoining wall.

Keeping these issues in mind, let's add some circles to the walls of our cup:

1. Orbit your model so that you can see inside of one of the walls (something like what is shown in *Figure 5.10*):

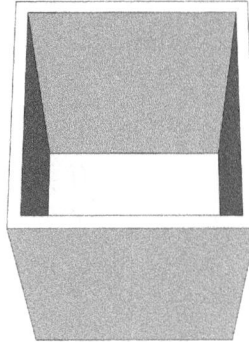

Figure 5.10 – Looking at the inside of a wall

2. Click **Circle**.
3. Draw a few circles out on the inside face of the wall.

    Keep in mind that, as you start the circle, your first mouse click is establishing the center of the circle, so make sure that your initial click is not too near the edge of the face or the bottom of the cup. Add a few circles and create something like what is shown in *Figure 5.11*:

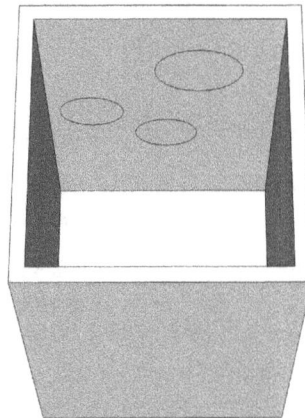

Figure 5.11 – Circles drawn on the inside face of the cup

Now, let's go ahead and do the same for the other three sides, making sure that the circles do not touch or extend past the inside faces.

4.   Orbit around the cup and use **Circle** to add circles to all the sides.

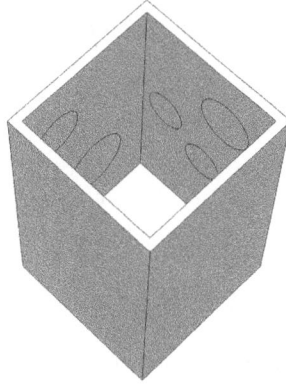

Figure 5.12 – All four sides have circles on the inside faces

5.   Use **Push/Pull** to push one of the circles to the outside face, creating a hole.

Once you have pushed one circle to the outside face and it disappears, you can quickly double-click on the remaining holes, and they will offset the exact same distance.

6.   Use **Push/Pull** to double-click on the remaining circles.

7.   Save your model.

This should give you the final geometry for your pencil cup.

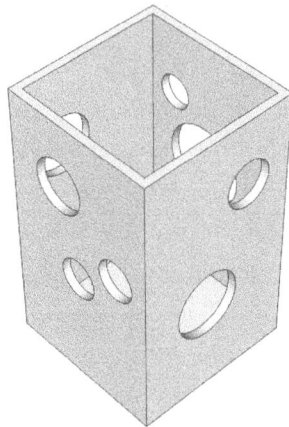

Figure 5.13 – The final pencil cup geometry

We kept this model fairly simple, but you can probably see how it is possible to create geometry for 3D printing using only the native tools in SketchUp. Our geometry is complete, but we are not ready to export our `.stl` file just yet. Next, we will make sure that we have a solid and get prepared to export!

Preparing your model for 3D printing

Right now, there are at least two things preventing us from calling this pencil cup print-ready. The first issue is that our geometry is not grouped. The second issue is the extra geometry in our file. Let's address these issues one at a time!

Making a solid group

We have already gone through what groups are and why they are important, and even how to use **Entity Info** to verify that they are solids, back in *Chapter 4, Print-Ready Modeling and Scaling for Export*. Let's get started by putting this geometry into a new group and then verifying that it is a solid:

1. Use **Select** to triple-click the pencil cup.
2. Right-click the selected geometry.
3. Choose **Make Group** from the context menu.

    Now, our geometry is in a group. The next thing to do is to verify that the group is a solid.

4. Open **Entity Info**.

Assuming you were able to follow every step listed as written, you should see **ENTITY INFO** reporting this as a **Solid Group** entity.

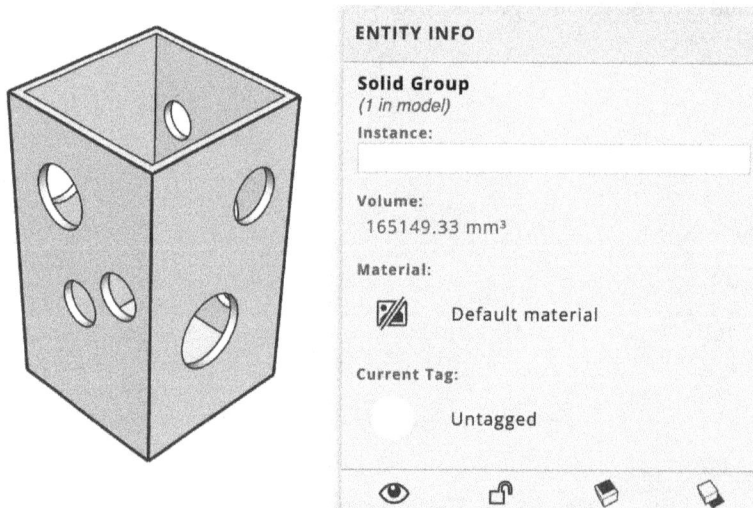

Figure 5.14 – ENTITY INFO showing that the pencil cup is a solid

This is great, and in a perfect world, this is how all modeling would go – create some geometry that is a solid, export it, and print it. Unfortunately, more often than not, there is some cleanup that needs to happen at this point. There is even a possibility that there is an issue in your model right now that is preventing it from being a solid!

The ability to diagnose issues with solids and being able to fix them is what separates the pros from the newbies. The skills needed to troubleshoot solids are especially important if you are using the Free version of SketchUp for Web. As we move forward in this book, we will see how **Solid Inspector** and other features of the paid version will help us to diagnose and fix problems with models. Right now, though, we are looking at 3D modeling in its purest form – just us and the geometry.

Assuming that everything did go to plan, and you are currently looking at a solid model, give yourself a high-five (yes, a self-high-five is just a single clap, but go ahead and clap for yourself – you deserve it), and read on. If your group is not currently a solid, then this next section should help you to shed some light on potential issues.

Diagnosing non-solid models

If your model is not a solid when you believe it should be, it means that you are most likely experiencing one or more of the issues we talked about in *Chapter 4, Print-Ready Modeling and Scaling for Export*. While it is great to know that issues may exist in a model somewhere, it is a whole new skill to know how to find and fix them.

For this section, we will be using the sample file for this chapter, 3D Printing with SketchUp Chapter 5 - Solids.skp. If you have not already downloaded this file, you can search for it on 3D Warehouse or use this link: https://3dwarehouse.sketchup.com/model/e8362510-9421-49a1-bbe3-fa19405c0373/3D-Printing-with-SketchUp-Chapter-5-Solids.

Let's open the sample file and troubleshoot some solids. If you still have your pencil cup model open, return to the **Home** screen and open 3D Printing with SketchUp Chapter 5 - Solids.skp. SketchUp for Web will automatically save your model when you return to the **Home** screen.

Example 1 – missing faces

For this example, you will need to expand both the **Entity Info** and **Scenes** panels. You should start in the **Cup 1** scene. If not, go ahead and click on the **Cup 1** scene now. Next, follow these steps:

1. Use **Select** to highlight the pencil cup on the screen.

    Note that **Entity Info** panel reports this as a **Group** entity (rather than a **Solid Group** entity). The first step I would take in troubleshooting a solid like this is to just look at the model. Spin the model around in 3D space and see whether you can catch a glimpse of what the problem might be.

2. Orbit around and look at all sides of the pencil cup.

    As you spin this model around, it should be quite apparent where the issue lies. The entire bottom of the cup is missing!

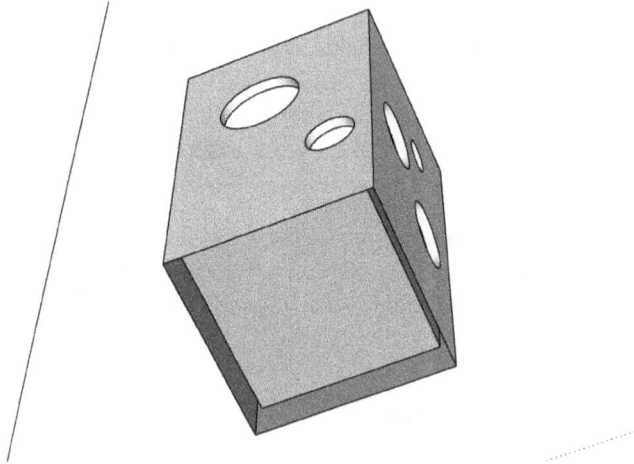

Figure 5.15 – The missing face on the underside of this model

This may feel like a silly example, but I can speak from experience when I say that this does happen. Sometimes, we get so wrapped up in modeling the top of something that we forget to look at the underside of our model!

The good news is that these sorts of issues are pretty easy to repair.

Advantages of the default material

In SketchUp, you can use the **Paint Bucket** tool to color faces whatever color you like. When I model for 3D printing, I tend to stick with the default material. While I enjoy color, I do this because it is a two-sided material. The front is bright white while the background is a blue-gray color. This color contrast is especially important when it is time to troubleshoot solids. If the front and back of the faces were the same color, seeing that there is a face missing could be a much more difficult process.

---

3. Double-click the group with **Select** to enter the group.

4. Use **Line** to draw over one edge of the bottom face.

5. Use **Select** to click outside the group to close it, and then click on the group again to highlight it.

At this point, **Entity Info** should report that the group is solid. It's really nice when your only issue with a group is a big missing face.

Example 2 – extra edges

An entire missing face is an easy problem to identify. Often, it will take a little more sleuthing. Let's head to the next model!

1. In the **Scenes** panel, click **Cup 2**, and then use **Select** to highlight the group.

    Entity Info is reporting that this is a group and not a solid. If you spin this cup around with **Orbit**, it looks the same as the cup we fixed in the *Example 1 – missing faces* section. Often, the reason that a model is not a solid is because of something that is going on inside the model.

    There are multiple ways to peek inside a model. You can use **Section**, or delete an exterior face and then replace it afterward. You could even try to use **Zoom** to try to poke the camera just under the surface of the outside face. My preferred method, however, is to temporarily hide a face and then put it back once we have fixed the problem.

2. Use **Select** to double-click into the group, and then select the front face of the cup.

3. Right-click on the highlighted face and choose **Hide** from the context menu.

    From here, you can see the inside of the geometry. If you look toward the bottom of the cup, you can see that there are extra edges connecting the bottom of the interior faces to the bottom of the cup. We need to get rid of these extra lines:

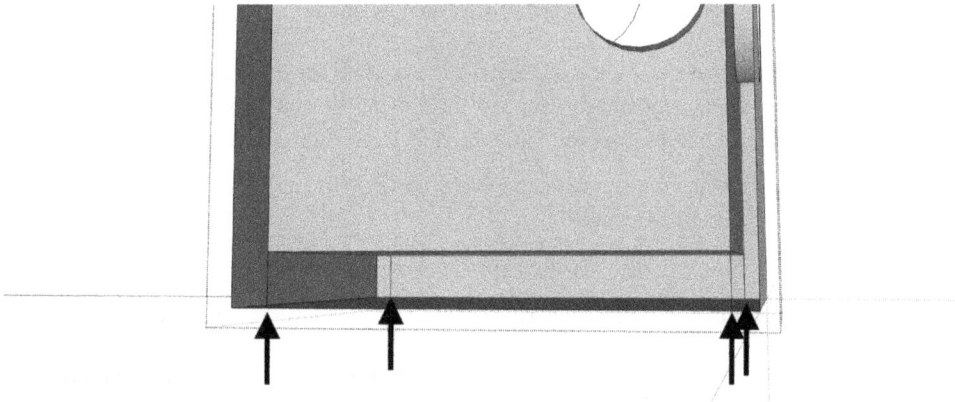

Figure 5.16 – Four extra edges that prevent this model from being a solid

4. Use **Eraser** to carefully erase these four extra edges.

5. In the **Display** panel, click the **All** icon under **Unhide** to bring the hidden face back.

6. Use **Select** to click outside the group to close it, and then click on the group again to highlight it.

Again, **Entity Info** should report that the group is solid. Often, edges like the ones in this example are left over from modeling and can be a bit tricky to find. Keep an eye out for places where **Push/Pull** was used to create new geometry to help find these issues.

Example 3 – extra faces

Sometimes, a whole face can get left inside your model. Let's take a look at the next example!

1. In the **Scenes** panel, click **Cup 3**, and then use **Select** to highlight the group.

    According to **Entity Info**, this cup is also not a solid. If you orbit around the model, you may find the issue quickly! At the back of the cup, you can see a face popping out of one of the sides.

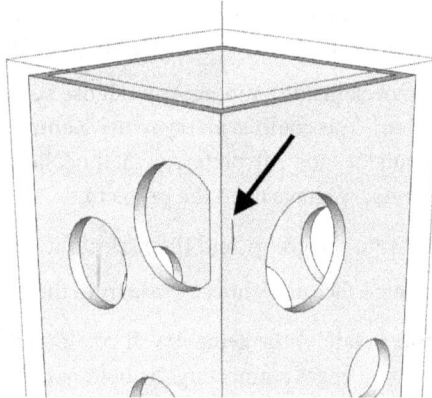

Figure 5.17 – An extra face inside the model

If you use the same process of hiding a face, you can see that the issue was that the circle used to create the hole was too big and overlapped the side of the cup. This was the reason that I suggested keeping your circles from touching the side walls when you drew them on the inside.

In this case, cleanup is as simple as erasing the extra face. If you have hidden a face to peek inside, use **Undo** to put it back now.

2. Use **Eraser** to erase the extra face, use **Select** to click outside the group, and then select it.

3. Once again, **Entity Info** should be reporting that this group is now a **Solid Group** entity.

While it would be impossible to run through examples of every single issue that can prevent geometry from being a solid, these are some of the most common, and there are some good methods for detecting issues manually. Now that we have finished with these examples, let's return to our model and get it ready to export!

Removing extra geometry

Let's open your `Cheese Cup.skp` file again and get it ready to send out to your slicer.

Remember, in the Free version of SketchUp for Web, everything in the model is exported as part of your `.stl` file. Let's verify what exactly is in your model by using **Zoom Extents** to see everything.

Figure 5.18 – Everything in our model

Unless you removed it at the beginning of the chapter, there should be a component of a scale figure towering over your little pencil cup. You will need to remove this from your model, along with any other geometry other than your pencil cup group, before going through the process of exporting a `.stl` file.

Exporting and printing

Since we spent an entire chapter (*Chapter 3, Importing and Exporting .stl Files*) going through the steps to export a `.stl` file, we won't go through them again here. If you like, though, please go ahead and export the file, and try slicing and printing this model. I did, and I now have a very nice-looking, homemade pencil cup holding my stuff!

Figure 5.19 – The final printed pencil cup

I do love how 3D printing lets me take an idea and turn it into something real in such a short amount of time!

Summary

In this chapter, we used basic modeling tools to create a printable solid geometry. We saw how to use those basic editing tools to put some holes in our solid geometry, and we even walked through a few examples of the kinds of issues you may run into when trying to model solids in SketchUp.

If you followed along with all the steps in this chapter, then you, too, should now be the proud owner of a custom, one-of-a-kind, SketchUp-designed, cheese-themed pencil cup.

Speaking of solids, in the next chapter, *Chapter 6, Modeling Using Solid Tools*, we will use SketchUp's **Solid Tools** to create another printable solid – this time from a photo!

6
Modeling Using Solid Tools

We have spent a lot of time talking about keeping our models solid as we create geometry. Anyone familiar with SketchUp has probably asked, why not use **Solid Tools** to create print-ready models? This is a great point and something that we will do in this very chapter!

For those unfamiliar with Solid Tools, they are a set of tools available in any paid version of SketchUp. Solid Tools allow you to perform basic Boolean operations on solid groups. These are functions such as adding solid groups together, subtracting one from another, or finding the volume where two solids overlap. They are great tools for working with solid geometry, thus a great set of tools to have in your 3D printing toolkit.

In this chapter, we will cover these main topics:

- Modeling from a photo
- Using Solid Tools
- Making solid 3D-printed connections

Technical requirements

For this chapter, we will be using Solid Tools in SketchUp for Web. This means access to SketchUp for Web with a SketchUp Go or Pro subscription. You will also need an image to model from. To follow along with a picture of your own or someone else's, a head in profile will work best.

Additionally, there is an example file for this chapter, `3D Printing with SketchUp Chapter 6 - Trophy.skp`, available at `https://3dwarehouse.sketchup.com/model/1151463b-05c5-44d5-9b03-67165bcb8f93/3D-Printing-with-SketchUp-Chapter-6-Trophy`.

This example file also includes finished versions of some of the models that are walked through in this chapter.

Modeling from a photo

For this example, we are going to start a model from scratch. I like the idea of making something that is fun to look at but still functional. Since one of the things that 3D printer owners seem to be asked to help with on a regular basis is creating awards (I have been asked to make trophies at least a half dozen times), we are going to be modeling a little trophy that you can display on your desk, something similar to what is shown in *Figure 6.1*:

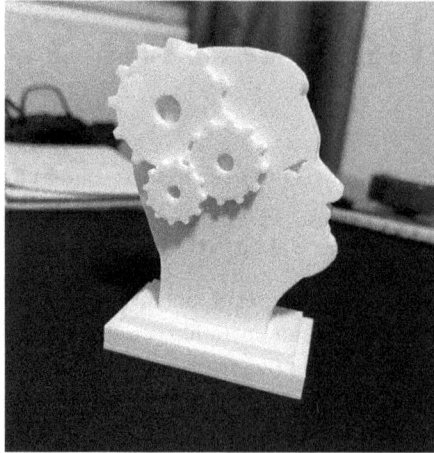

Figure 6.1 – Final printed trophy

I say that it will look something like this because you will be using your own image for this model. If you do not have a photo to work from, you can copy the image I used from the example file, 3D Printing with SketchUp Chapter 6 - Trophy.skp, in the scene named **Reference**.

With that, let's hop into SketchUp for Web and model a trophy! In this chapter, I will be showing the SketchUp Go or SketchUp Pro version of SketchUp for Web. The paid version of SketchUp for Web includes additional tools (namely, Solid Tools and Solid Inspector) shown throughout this walkthrough. With that, let's open SketchUp for Web in your browser and start modeling!

Outlining the head

The first thing we will create is an outline of the head that we can use to create the main portion of the trophy. To do this, follow these steps:

1. Click **Create New** to start a new model.
2. Click the menu icon, then click **Import**, followed by **My Device**.
3. Click **My Device**, navigate to your head photo, select it, and click **Open**.
4. Click to place the image on the ground. Make sure to make it fairly large, as shown here:

Figure 6.2 – Profile image imported into SketchUp

Large-scale modeling

When modeling something that has a lot of curves or modeling geometry that you plan to use with Solid Tools, it is a good idea to model at a larger scale than you intend to print. As with the rest of SketchUp, Solid Tools were created with the intention of working on architecture-sized models. If the geometry gets too small, Solid Tools can create unreliable results. In this case, we are going to start out modeling very big, then we will scale down once we have the geometry we want to export.

5. Use the drawing tools to trace your profile.

    How you do this is up to you. You can use the **Line** tool to connect a string of segments or the **Arc** tool to create a series of curved lines. I used the **Freehand** tool to trace most of the face, then the **Line** tool to draw in the neck as straight lines. The important part is that it closes and creates a face, covering your face.

    Also important is the "neck" of the shape. Make sure that you add geometry that extends down from the neck and ends in a nice rectangle, like what is shown in *Figure 6.3*:

Figure 6.3 – A single surface created to cover the face in the photo

6. Use **Select** to triple-click the face, then hold down the *Shift* key, and click on the reference image.

7. Use **Rotate** to rotate the selected geometry so that the line at the bottom of the neck is parallel to the red axis, as shown in *Figure 6.4*:

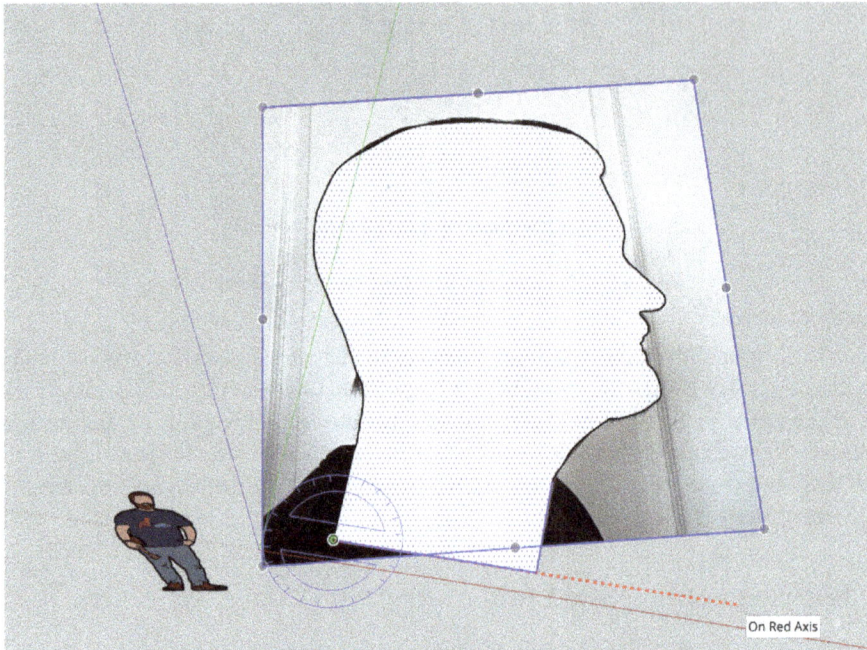

Figure 6.4 – Geometry rotated to the red axis

This step is not strictly required, but it will make some of the modeling ahead a little easier. It is very important to make sure that you rotate the head face and the reference geometry together when you rotate, as we will reference the photo in a future step and so need to keep it aligned with the outline geometry we have created.

8. Now, use **Push/Pull** to pull the face up into 3D.

9. Use **Select** to triple-click the geometry, then right-click on it, and choose **Make Group** from the context menu.

This should give us our first solid group to work with! Here's how it should look:

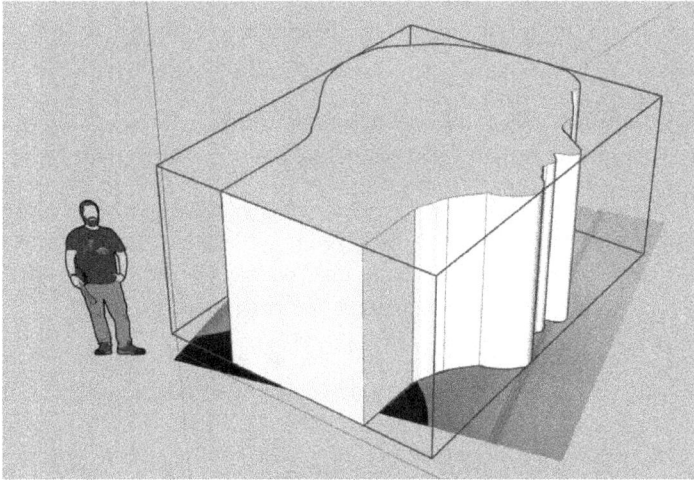

Figure 6.5 – A solid head group

10. On the right side of the screen, click the very bottom icon and open the **Solid Inspector** panel. Let's take a quick look at this feature of the Go or Pro version of SketchUp for Web.

While using the **Entity Info** panel is a quick and easy way to check whether something is a solid, it is not the best tool for the job. Let's take a short break from modeling and take a look at how Solid Inspector can help you get better at creating solid models.

Using Solid Inspector

Solid Inspector will display information about the currently selected group (and only one group) and let you know whether there are any issues. With Solid Inspector, two different kinds of issues may be identified. Some errors can be automatically fixed while others can be identified, but you—the user—will need to fix the issues yourself.

Errors that can be fixed automatically are as follows:

- **Reversed faces**—These are faces where the outside face is facing inward.
- **Stray edges**—These are extra edges that do not connect faces together.
- **Internal and external faces**—These are faces that are not connected to the main solid geometry. These extra faces may be inside or outside of the solid shape.
- **Face holes**—A face hole is just that: a hole in a face.

When you run Solid Inspector and get one or more of these types of errors, you will have the option to automatically fix the issues. You can, of course, fix them yourself, but in the majority of cases, Solid Inspector will do a good job of addressing the issue.

The second kind of issue Solid Inspector may find are issues that you might need to fix yourself. Those errors are as follows:

- **Border holes**—These are holes at the edge of a mesh. Unlike holes in a face, which can be fixed automatically, holes on the border of your model are not always straightforward to repair. These holes need to be fixed manually.

- **Nested groups/components**—This error lets you know that you have groups or components inside the group or component you are inspecting. You may need to explode the internal objects or remove them from the group or component that you are working on.

- **Image entities**—This message lets you know that an image (such as the one we imported in the *Outlining the head* section) exists inside the group or component being inspected. Images cannot be made solid as they are single faces, so you will need to delete them to get a working solid.

- **Short edges**—If your model has very short edges, you may see this message. In some cases, very small geometry can cause issues with exporting and printing. Technically speaking, your model can be solid, export, and print just fine, even though Solid Inspector has identified short edges. In many cases, replacing short edges with longer edges would mean recreating some portion of your model. I generally take note of this issue but do not change my model if this is the only issue that Solid Inspector reveals.

If you run Solid Inspector and it does find errors, you can click on the error message, then use the buttons in the panel to navigate through the errors and, if the error can be automatically fixed, click the **Fix Errors** button.

Creating an intersecting shape

Next, we need to create a second solid that will give us a rounded shape—something that will give our head shape some depth without having to sculpt a bunch of complicated geometry. Once we have that geometry created, we will use Solid Tools to merge that shape with the geometry in our head group.

Off to the side of our existing model, we are going to create a shape like the one shown in *Figure 6.6*:

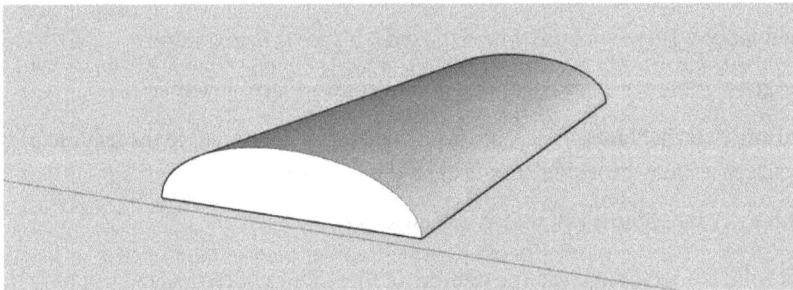

Figure 6.6 – The second solid shape for our model

How many sides in a circle?

There are more than a few schools of thought when it comes to how many sides you should use when drawing a circle in SketchUp. For most models, whatever looks good based on how large the circle is works. When creating geometry that will be 3D-printed, you need to consider how large the final circle will be and whether the sides of the circle will show in the final model. I generally model with a few rules of thumb. If the circle will be small (such as a screw hole or perforation under an inch), I will use 24 sides. If the circle is going to be the entire model (if I were modeling a plate or a wheel), I use 192 sides. Anything that falls between will get 96 sides. While this standard can create unnecessarily heavy models for other purposes, I find that it works well for 3D printing.

Let's create this squat half-circle, a new shape, and make sure it is solid:

1. Start the **Circle** tool.

2. Immediately set the number of sides on the circle by typing 96.

3. Lock the circle perpendicular to the green axis by pressing the *left arrow* key.

4. Draw a circle that is approximately wider than your head group.

    Next, we need to cut the circle in half.

5. Use **Line** to draw a line from the center of the circle, along the red axis to the right edge, then a second line from the center to the left.

6. Use **Eraser** to remove the bottom half of the circle.

    Now, let's pull this half-circle into a 3D shape.

7. Use **Push/Pull** to drag the shape out, approximately longer than your head group.

8. Use **Select** to triple-click the shape, then click on the **Scale** tool.

9. Drag the top center handle downward, creating geometry like what we see in *Figure 6.6*.

10. Finally, use **Select** to triple-click to select all the geometry, then right-click, and select **Make Group** from the context menu.

11. Use **Solid Inspector** to verify that this is a solid.

Now, we can look at using Solid Tools to create a new solid from the two solid groups.

Lining up the groups

The last step in this section is to align our two shapes. We need the head shape and the bar we just created to overlap so that the head extends past the bar vertically, but the bar laps just past the face on all other sides.

Use the **Move** tool to move the bar group so that it laps over the face group, like what is shown in *Figure 6.7*:

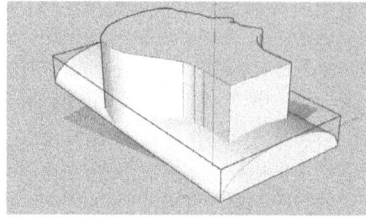

Figure 6.7 – Overlapping groups, ready for Solid Tools

If the bar is not big enough to cover the head group, use **Scale** to make it a little bigger.

This gives us a couple of groups that SketchUp recognizes as solids. Up next, we will see how we can use those shapes with Solid Tools to generate some unique geometry!

Using Solid Tools

Solid Tools are great tools for interacting with simple and complex 3D shapes. Using Solid Tools, you can cause two or more solid groups to combine or use one solid as a way to subtract from another. You can even use Solid Tools to create a new volume that represents where two shapes overlap in 3D space!

> **Groups, components, and Solid Tools**
>
> The decision to use groups or components can have many reasons behind it. In the case of Solid Tools, you will want to use groups. Once you perform an action using Solid Tools, the resulting geometry will always be created in a new group. This means that, even if the initial geometry is in components, you will end up with a group. For this reason, I usually stick with groups initially and make them into components after I use Solid Tools, as needed.

All of the Solid Tools can be accessed by clicking on the active Solid Tool on the toolbar (by default, this is the **Outer Shell** button, but if you have used a different Solid Tool recently, the icon for that tool will be displayed):

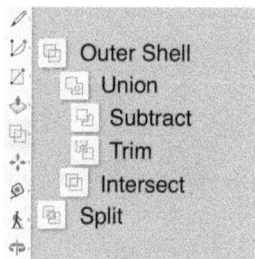

Figure 6.8 – All Solid Tools

Before we start using any of the Solid Tools, let's do a quick overview of what each one does, and the order in which you want to pick your solids:

- **Outer Shell**—This Solid Tool will combine any number of solid groups into a single solid group consisting of only the exterior faces from the original geometry. If there are any voids or open spaces between the selected groups, they will be filled in when using **Outer Shell**. To use this command, you need to have two or more solid groups selected, then click on the command.

- **Union**—This Solid Tool will allow you to combine two solid groups into a single solid group. Unlike **Outer Shell**, **Union** will allow you to create a solid group with empty space inside. To use this command, select one group, then click the **Union** icon, and then click on a second solid group.

- **Subtract**—This Solid Tool will allow you to remove the volume of one solid from another. To use this command, you will start by selecting the solid group that you want to remove, then you will click the **Subtract** icon, and then you will select the solid group that you want to subtract from. The result will be a new solid group that has the geometry from the second group, minus the geometry of the first group. The first group is completely removed from the model.

- **Trim**—This Solid Tool works similarly to **Subtract** but keeps the subtracting group in the model. To use this command, start by selecting your subtracting tool, then click the **Trim** icon, then select the group that you want to remove geometry from. You will end up with a hole in the second group that the first group fits perfectly into.

- **Intersect**—This Solid Tool will find the geometry where two solid groups overlap and remove everything else. To use this command, select one solid group, then click the **Intersect** icon, then click on the second, overlapping solid group.

- **Split**—This Solid Tool works like **Intersect** but will maintain everything outside of the overlapping geometry. To use this command, select one solid group, then click the **Intersect** icon, then click on the second overlapping solid group. The result will be three separate solid groups.

Let's hop in and use some of these commands on our trophy model—specifically, **Intersect**, **Subtract**, **Union**, and **Trim**.

Trimming two groups

We will start by removing the extra geometry from our overlapping groups using the **Intersect** Solid Tool:

1. Use **Select** to highlight the bar group.
2. Choose **Intersect** from the toolbar.
3. Click on the head group.

This will create a new shape made up of the volume where both groups completely overlap:

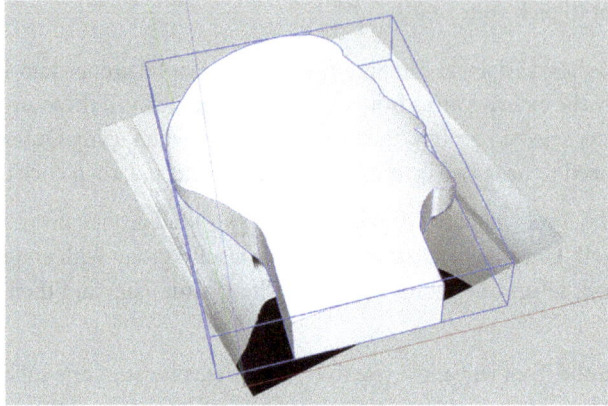

Figure 6.9 – Solid group from Intersect

Let's keep going and use a few more Solid Tools!

Carving out details

Next, we are going to add a little bit more detail by adding a void to the head matching the eye in the photo. The goal is to make this little trophy identifiable as the person from the picture without having to add a lot of detail, and the shape of the eye does just that.

To add more details, we will follow these steps:

1. Use **Select** to highlight our new head group, right-click, and choose **Hide** from the context menu.

2. Using **Lines**, **Arcs**, or **Freehand**, trace the shape of the eye on the image:

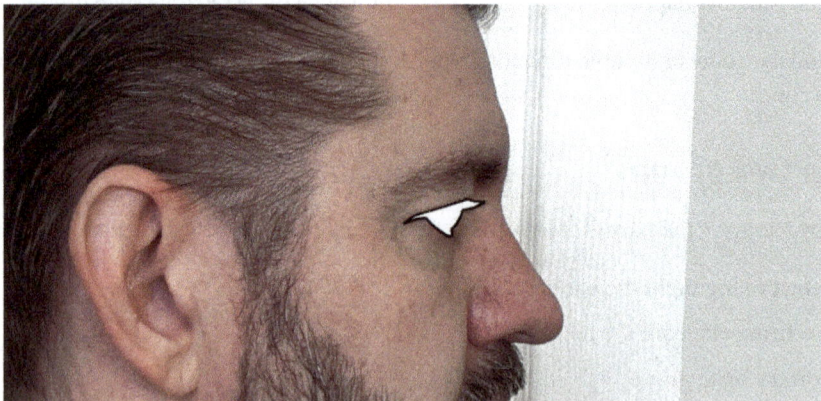

Figure 6.10 – Lines used to trace the eye in the image

This will end up being very small in the final print, so you do not need a ton of detail—just enough that you can identify it as the eye of the person in the image.

Now that we have a face to work from, we can go about making a solid.

3. Use **Push/Pull** to pull the face up into a 3D shape.

4. Use **Select** to triple-click the geometry, then right-click, and choose **Make Group** from the context menu.

    Now, before we use any Solid Tools (you should verify that your new shape is a solid group in **Entity Info**) on this group, let's lift the group up, off the ground a bit. We want the eye to be a void in the final head volume, but not a hole that goes all the way through.

5. Use **Move** and the *up arrow* to move the group vertically so that it is away from the image.

    Now, let's get our head group back so that we can subtract the eye group.

6. In the **Display** panel, click the **All** icon in the **Unhide** section of the panel.

    At this point, you should see your head group with an eye-shaped stick in its eye, something like what is shown in *Figure 6.11*:

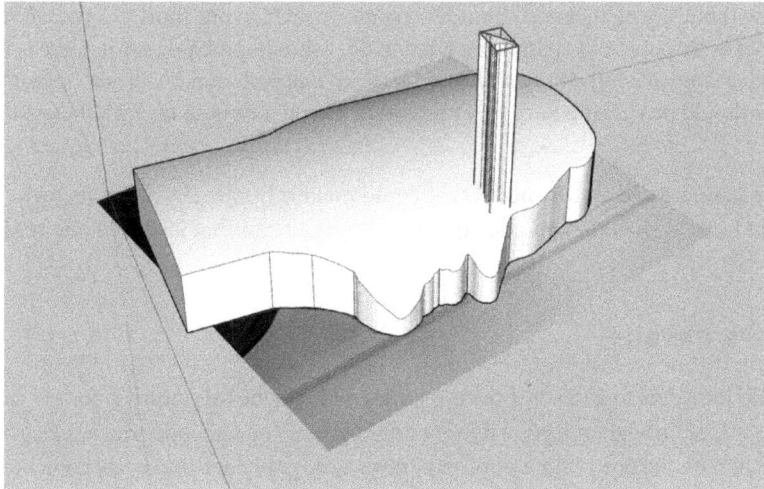

Figure 6.11 – Two solid groups, ready for Subtract

7. Use **Select** to highlight the eye group, then choose **Subtract** from the toolbar, then click on the head group.

    Assuming both groups were still solids (and you know what to do if they were not), this would result in a single solid group that looks just like your head group but with an eye-shaped hole where the eye should be!

Figure 6.12 – Head solid with eye shape successfully removed

8.   Use **Eraser** to get rid of our reference photo, as we will not need it anymore.

Saving your model

It should go without saying that you should be regularly saving your model as you go through these steps, but as this is a book telling you what to do, I should probably say it. Here is my rule for when to save: *any time you do something in SketchUp that you would not want to do again*. To that end, you should be clicking **Save** every few steps as you follow along with these examples.

At this point, we have created a little model that could make a fun trophy all on its own. I think we should go a little farther, though, and add something to the head that indicates the reason for the trophy being given.

Adding more details

I thought it would look cool to add some overlapping gears. Maybe this award is for someone who is great at thinking things through or a good designer or engineer (or someone who is a gearhead, if you don't mind being too on the nose with it). So, let's draw a gear group and add a few copies to our head:

1.   Pan off to the side of the head group.

2.   Use **Circle** to draw a circle with 24 sides on the ground, making sure that you pull the edge out on the green or red axis.

     We just covered some rules of thumb on how many sides a circle should have when modeling for 3D printing. Since this circle will just be geometry used to create a gear, there is no need to add a bunch of extra sides. In fact, it will be easier to create the gear with fewer sides.

3.   While still in the **Circle** tool, add another, smaller circle to the center of the larger circle:

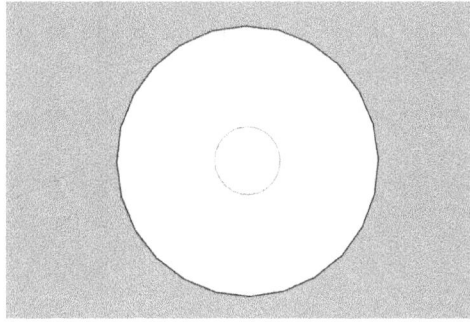

Figure 6.13 – Two circles, ready to become a gear

Next, we will draw a single tooth, which we can copy in an array around the circle. Now, my aim here is a gear-shaped solid that we can merge with the head. For this purpose, I will not be worrying about the specific math behind creating a working gear and the proper tooth size, spacing, or depth.

We will be drawing three lines with the goal of creating a shape at the top of the circle that looks like the geometry shown in *Figure 6.14*:

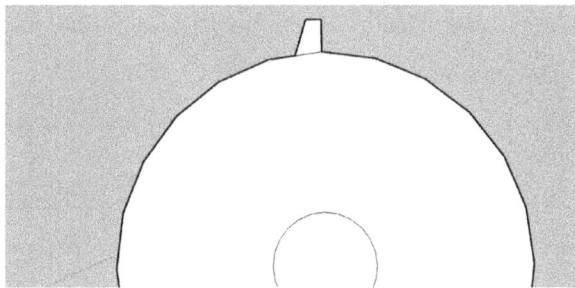

Figure 6.14 – Half of a single tooth

4. Use the **Line** tool to draw a line up from the very top of the circle.

5. Draw a second line (about half the length of the first) straight to the left.

6. Now, draw a third line from the end of the second to the midpoint of the side of the circle below.

    Now, we just need to copy that geometry to the other side so that we have a symmetric tooth. There are several ways to do this, but I find the easiest to be using the **Rotate** tool.

7. Use **Select** to highlight the half-tooth face that we just created.

8. Click the **Rotate** command and then press the *left arrow* key (this will lock the rotation to the green axis) and the modifier key to toggle copy (*Ctrl* on Windows and *Option* on Mac).

9. Click once on the point at the top of the circle.

10. Move your cursor to the left, and click a second time.

11. Now, move your cursor to the right and click a third time (the **Angle** setting in the lower right of the screen should read **180**).

This should copy the tooth face to the right of the original, giving you the geometry of a full tooth, as shown here:

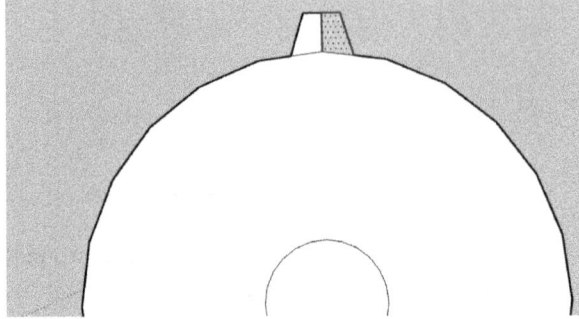

Figure 6.15 – Copied geometry creates a symmetrical tooth

12. Use **Eraser** to remove the vertical line in the middle of the tooth.

Now, we just need to duplicate that tooth all around the circle. We will do this with **Rotate**, as well.

13. Use **Select** to highlight the tooth face.

14. Click **Rotate** and click the center of the circle (you may need to hover over the edge of the circle for just a second for the center-point inference to appear).

15. Now, click on the point at the top of the circle (the middle of the selected tooth).

16. Finally, click on the circle two points to the right of the center (check out *Figure 6.16* if you need clarification):

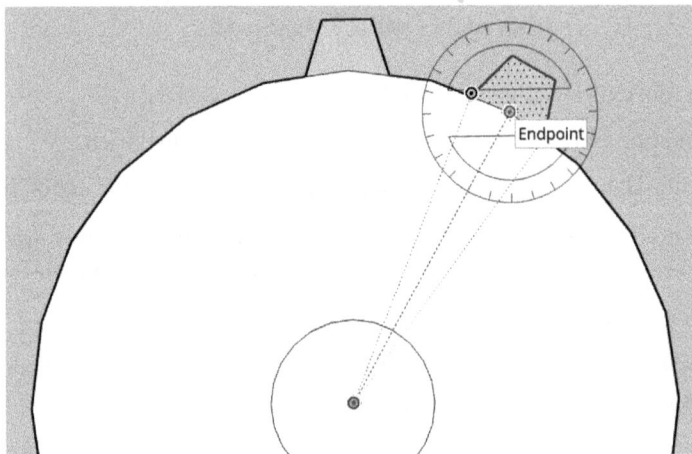

Figure 6.16 – One tooth copied

17. Type `11x` and press the *Enter* key.

    This will give us a total of 12 teeth (the original plus 11 copies). It is important that you type the number of copies you want immediately after the first copy; otherwise, you will have to start the process over.

    The final step to finishing our gear is to merge all these shapes into one and make it a solid group.

18. Use **Eraser** to remove the lines that separate the teeth from the circle.

19. Use **Select** to highlight the circle in the middle of the gear, then press *Delete* to remove it.

20. Use **Push/Pull** to pull the gear face into 3D.

21. Use **Select** to triple-click the geometry, right-click, and choose **Make Group** from the context menu.

    At this point, you should have two solid groups in your model—one of your head model and the other a 3D gear:

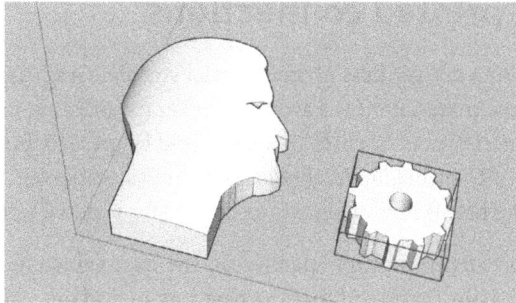

Figure 6.17 – Head group and gear group

22. Using **Move** and **Scale**, create a few copies of the gears of different sizes and arrange them on the upper-left part of the head.

    This part allows you to be creative. Overlap the parts and make the gears of different sizes. Make sure that the gears are flat to the ground plane and extend up past the top of the head group (I used **Scale** to make the gears taller as needed). I ended up with three different-sized gears:

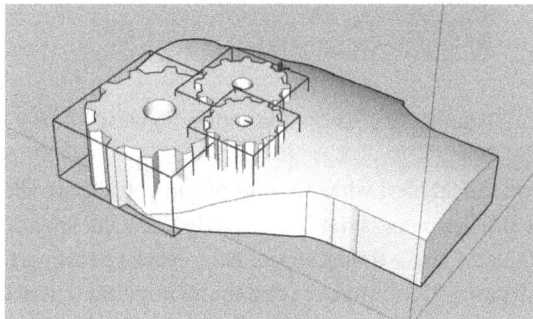

Figure 6.18 – Gear groups overlapping the head group

The final step now is to merge them all into one solid group.

23. Use **Select** to highlight all the gears and the head group.

24. Click the **Outer Shell** icon from the toolbar.

25. Use **Rotate** to tip the head group so that it stands up vertically.

26. Click the **Scale** tool and press the *Shift* key.

27. Click on the handle in the middle of the top of the bounding box and start to move it down. Type 4 " and press the *Enter* key.

You now have the bulk of the trophy created and at the correct scale. The final piece to create is the base. Before we can do that, though, we need to talk about how printed pieces fit together, which we will do in the next section.

Making solid 3D-printed connections

When you print with a 3D printer, regardless of the material you are using, there will be swelling and shrinking. When printing, the material moves around during the printing process, and everything expands and contracts as materials cool or cure. While it would be great to think that a 1" 3D-printed peg will fit perfectly into a 1" 3D-printed hole, that is rarely how it works out. Generally, you need to allow for some additional space between printed parts to account for material expansion during printing.

I have heard multiple reports from multiple parties on the best gap to include in mechanically connected parts when printed but I've found that the best way to find out what kind of gap to use is best determined by testing your own printer. To that end, let's make this quick test model:

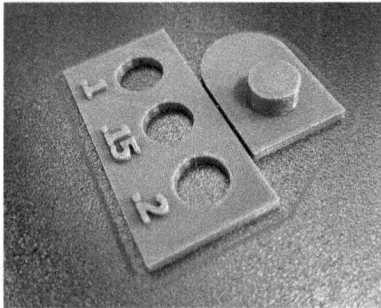

Figure 6.19 – 3D-printed connection test

With this simple test, you can quickly test what amount of gap you should allow for in your printed connections. The peg in the smaller piece is modeled exactly 1/2" while the holes in the larger piece are 0.1 mm, 0.15 mm, and 0.2 mm larger than the peg. After printing this part, you can slide the peg into each hole and see which gap is ideal for your printer and material. There is a copy of this model in the 3D Printing with SketchUp Chapter 6 - Trophy.skp file on the **FDM Test** scene. I recommend exporting this test and printing it before proceeding with the trophy model.

> **Materials matter**
>
> The test shown in *Figure 6.19* was designed for use with an FDM printer. If you are printing with an SLA printer, check out the test on the **SLA Test** scene. Since SLA printing creates a finer print, material moves less during the printing process, requiring smaller gaps between pieces. Thus, the test is very similar to the FDM test, but with smaller gaps.

Once we know the proper gap for pieces, we can move forward with making a base for our trophy. This is a simple shape that can be created in just a few steps:

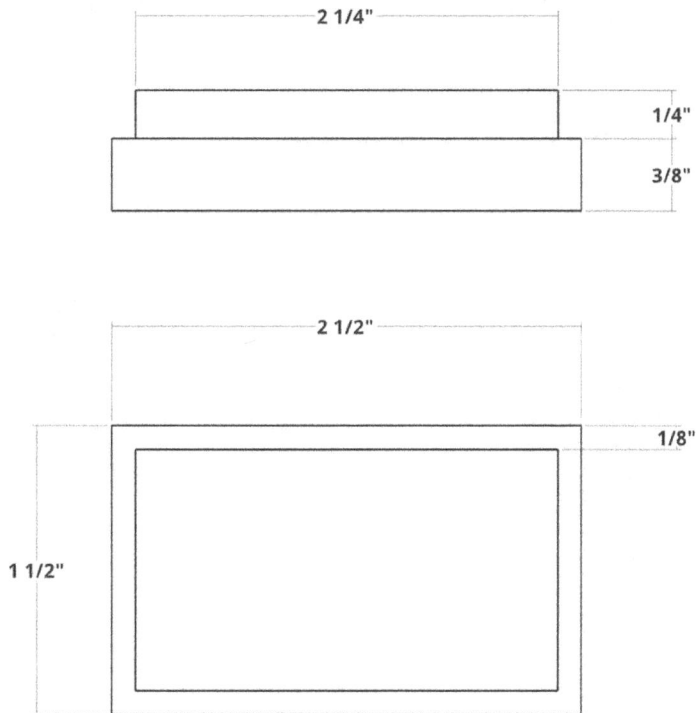

Figure 6.20 – Dimensions for the trophy base

Follow these steps to model the base shown in *Figure 6.20*:

1. Start by using the **Rectangle** tool to draw a rectangle that is 1 1/2" x 2 1/2".

2. Use **Push/Pull** to pull the rectangle up 3/8".

3. Use **Offset** to offset the outer edge of the top of the box at 1/8".

4. Use **Push/Pull** to pull the new rectangle up 1/4".

5. Use **Select** to triple-click the base geometry, then choose **Make Group** from the context menu.

With the geometry of the base created, the final step is to make a hole in the base that the head group can slot into.

6. Use **Move** to position the head group so that the elongated neck sits into the base.

    Since both pieces are currently flush to the ground plane, if we were to use Solid Tools to cut a hole in the base right now, it would go all the way through the base. Let's move the head group up just a bit.

7. Use **Move** to move the head group up the blue axis 1/4".

8. With the head group still highlighted, click the **Trim** icon from the toolbar, then click on the base group.

9. Use **Move** to slide the base group off to the side so that you can clearly see the hole in the top.

    All that is left to do is to add the gap to the hole, and we are ready to export some geometry.

10. Use **Offset** to offset the hole outward with the gap amount you liked best from the gap test (for my model, this was a 0.15 mm offset). It should look something like this:

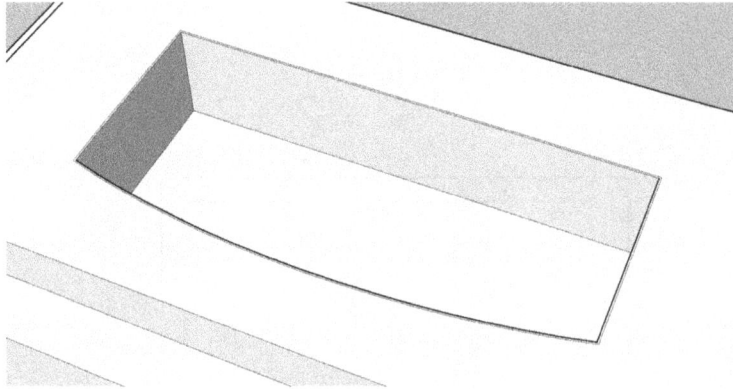

Figure 6.21 – Outline of the hole with a small offset

11. Use **Push/Pull** to push the offset geometry down to the bottom of the hole.

12. Use **Eraser** to remove the offset line from the bottom of the hole.

    The modeling is complete! You now have a finished award in two pieces that should connect perfectly once printed!

Before exporting, you will need to clean up your model, of course. Remove the scale figure from the model and choose how you want to export it. For my export, I chose to export both pieces—the head and the base—together. I laid both pieces right next to each other and exported a single .stl file:

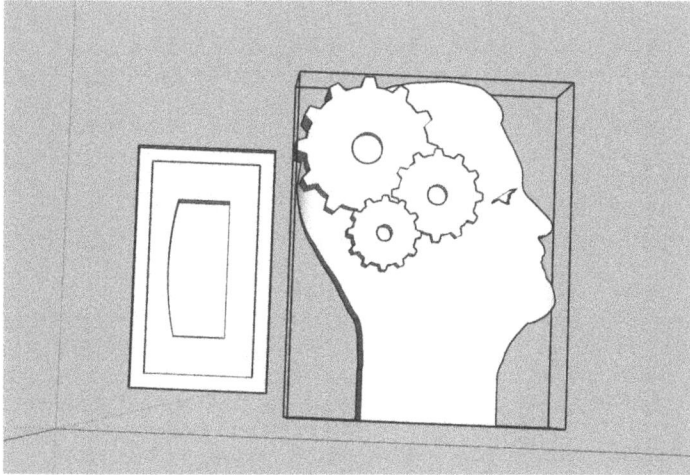

Figure 6.22 – Final model, ready for export

Remember—this is not just an example file that helps you understand how to use the tools in SketchUp to create and print a fun little model; it is also an achievement that you get to print and keep on your desk! Call it the *"Finished Chapter 6 of 3D Printing with SketchUp"* award!

Summary

We covered a lot in this chapter! We created a multi-piece model from a photo, and you also got to see how to use an image as a reference, dipped your toes into the world of Solid Tools, and learned how to account for gaps in the model to create snug post-print connections. Plus, you have a blueprint that you can use if you or anyone you know wants you to use your 3D printer to recognize a co-worker or friend!

Up next, let's talk about working with existing geometry in *Chapter 7, Importing and Modifying Existing 3D Models.*

Importing and Modifying Existing 3D Models

I am going to go out on a limb here and say that anyone who owns a 3D printer has printed a file that they did not create themselves at some point. Be it the sample file that comes with your printer or a really cool model downloaded from one of the many file-sharing sites that cater to 3D printing enthusiasts, the odds are high that you have printed something that you did not make from scratch. Sharing our models for 3D printing has become something of a standard and part of what we do as 3D printers. We download, modify, print, and re-share models all the time (remembering to give credit to the original creator, of course!).

In some cases, you may end up needing a file that you don't have the ability or time to create from scratch. It's also possible that you see a model and just think it would be fun to print! Whatever the case, there is always a possibility that you will need to make modifications to a downloaded file that goes beyond what your slicing software can do. If all you need to do is change the scale of the model before printing, your slicer should suffice, but if you want to go beyond that and add a geometry, or if the downloaded file has issues that prevent it from being printed, you may want to edit the geometry in SketchUp.

In this chapter, we will cover these main topics:

- Importing existing 3D models
- Repairing existing models
- Modifying existing models

Technical requirements

In this chapter, I will be using SketchUp Desktop, as well as the extension with Solid Inspector[2]. Additionally, we will work with the 3D Printing with SketchUp Chapter 7 - Importing. skp file, which you can find on 3D Warehouse: https://3dwarehouse.sketchup.com/ model/38ccf96a-e305-45f5-b09f-b9a211d6196e/3D-Printing-with-SketchUp-Chapter-7-Importing.

Importing existing 3D models

In this chapter, we will be working with a model imported from a .stl file. This file is of a cool-looking rhino head. We will use this model to make a coat hook that we can hang on the wall. Following the steps in this chapter, we will end up with a final useable model like the one shown in *Figure 7.1*:

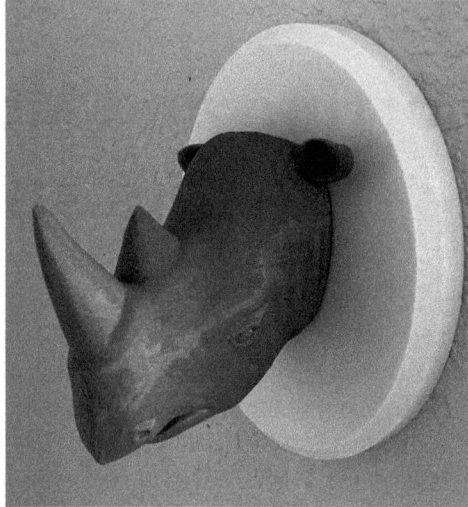

Figure 7.1 – Our final rhino head coat hanger

When you get your hands on a model intended for 3D printing, the odds are high that it will be in the .stl format. It is possible, however, that it comes to you in another format. Fortunately, SketchUp can import a number of the most common 3D geometry file formats. In addition to a .stl file, you can also use SketchUp Pro's **Import** command, with which you can import any of these file formats:

- **Digital Elevation Model (.dem)** – This is a file format most often used to represent a 3D geometry relative to the landscape geometry

- **Drawing exchange Format (.dxf)** – This is a non-proprietary file format often used by CAD software to pass geometries back and forth

- **Digital Assets Exchange (.dae)** – Often referred to as a COLLADA file, this is a format that passes 2D and 3D imagery back and forth between many platforms

- **Keyhole Markup Zip (.kmz)** – This is a compressed file that contains map information, including 3D geometry

- **3D Studio (.3ds)** – This is a 3D Studio Max file that contains 3D mesh geometry

- **Wavefront Object File (.obj)** – This file format contains basic 3D geometry data and is used by many 3D modeling tools as a basic file format for transferring points and faces in 3D space

While it would be interesting to import and examine each of these types of files, we will work from a file that has already been imported with our example file for the chapter, `3D Printing with SketchUp Chapter 7 - Importing.skp`. This file contains the results of importing a `.stl` file.

Extensions to consider

If you find yourself importing geometry files regularly, there are a few extensions that you may want to consider.

Skimp (`https://skimp4sketchup.com/`) – This tool will make importing models with dense geometry easier by allowing you to reduce the detail in the mesh so that it creates a SketchUp model, which is easier to work with.

Transmutr (`https://lindale.io/transmutr`) – This is another importer that will give you more control over your imported meshes.

CleanUp[3] (`https://extensions.sketchup.com/extension/046175e5-a87a-4254-9329-1accc37a5e21/clean-up`) – This extension will make removing extra edges from imported geometry quick and easy. This is a great way to keep file sizes smaller when working with large, imported geometries.

Let's open this example file in SketchUp for Desktop and take a look at what your next steps should be after importing geometry.

Examining your import

The first thing that I always do after a successful import (yeah, I have downloaded more than a few files that I could not get to import at all) is spin around and look at what I have going on. Start by simply looking at the model, orbit around, and see whether you see any major issues:

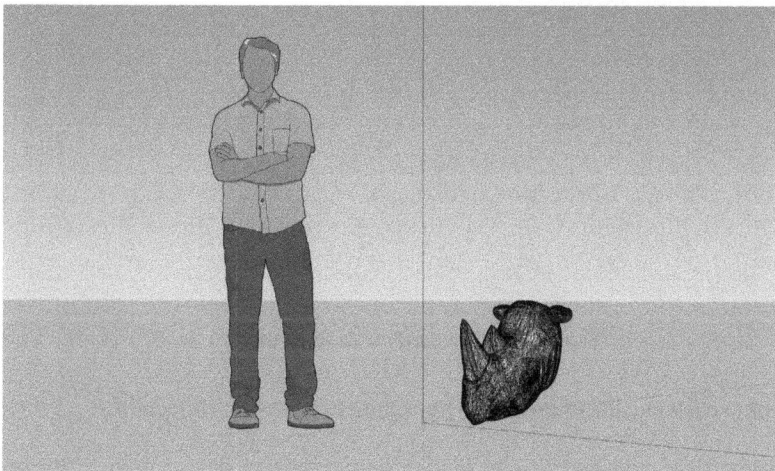

Figure 7.2 – Rhino head .stl file successfully imported

As I rotate around this model, there are a few things that I notice right away. First off, this thing is big. If I compare it to Niraj (the scale figure for SketchUp 2022), I can tell that this is a couple of feet tall. Second, if I flip around and take a look at the back side, I can see at least one hole in the mesh. There may be more holes, but it is hard to tell with all the edges visible here.

To make the model a little easier to examine, I may want to try smoothing the geometry. This may not be something that you need to do on all imports (an imported house, for example, might be all flat surfaces and may not benefit from being smoothed), but in this case, smoothing will make it easier to see the surface of the model. Before we troubleshoot any further, let's smooth our rhino head out:

1. Use **Select** to double-click on the rhino head group.

2. Triple-click on the rhino head mesh, then right-click, and select **Soften/Smooth Edges** from the context menu.

    If you don't already have the **Soften/Smooth Edges** panel open, this will open it for you.

3. Check the **Soften Coplanar** checkbox and move the slider so that most of the edges disappear.

    The goal is to get most of the edges to smooth out. Hard edges (such as the ones that create the flat back side of the model) should stay visible. Try moving the slider so that the majority of the edges are smooth but all the edges around that back face are still visible:

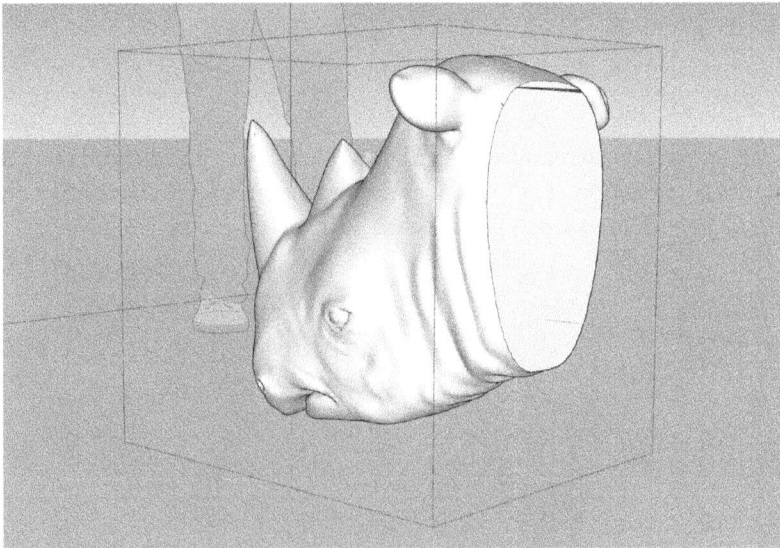

Figure 7.3 – Selected mesh smoothed

4. Use **Select** to click outside of the group bounding box to close the group.

As soon as you smooth the mesh and spin the model around, you may detect another issue or two in the model. There is more than one hole in this mesh, and we can actually see them with the edges smoothed.

What about the unsmoothed edges?

When we smoothed the mesh, we ended up with some edges (around the eye and under the ear, for example) that were not smoothed out. If you like, you can zoom in and use **Eraser** with the **Soften/Smooth** modifier to get rid of these edges. This is extra work that will not change the model for export, though. When we end up exporting this model for printing, all of the triangles in the model will export. The smoothing is just something that we do in SketchUp to help us visualize the surface and does not affect the mesh that is actually exported.

So, we know at this point that there are at least a few holes in this model. Let's take a closer look and use Solid Inspector² to find out exactly what we need to fix to make this model solid.

Checking your model with Solid Inspector²

Right now, if you select the group and look at your **Entity Info** panel, you will see that this is a **Group** object, rather than a **Solid Group** object, which is what we want. Let's fire up Solid Inspector² and see why this group is not solid:

1. Use **Select** to highlight the rhino head group.
2. Click on the **Tools** menu and then select **Solid Inspector²**.

    This will bring up the **Solid Inspector²** window and, after a few seconds, let you know what issues prevent the selected object from being solid:

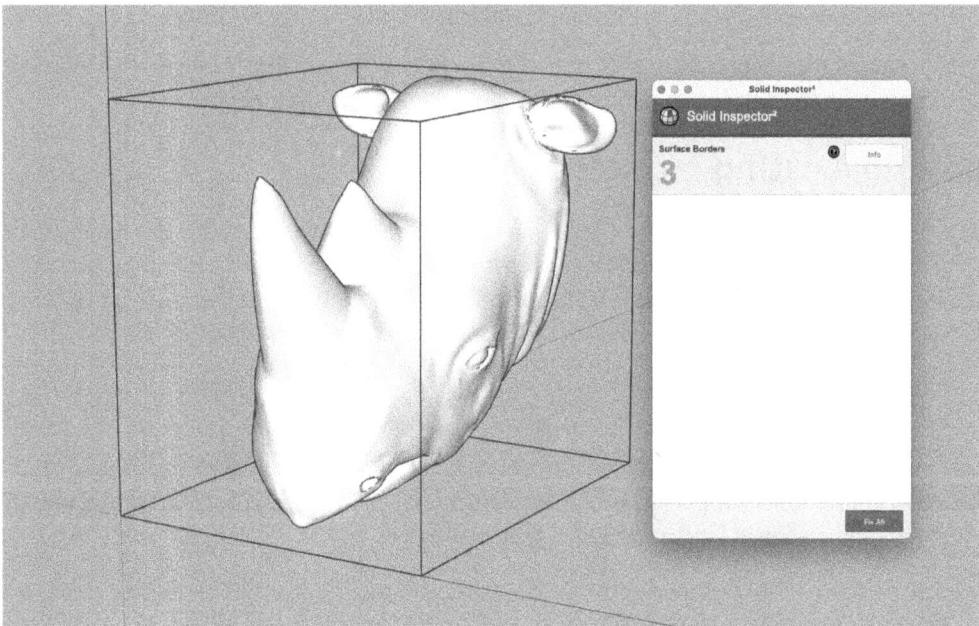

Figure 7.4 – The group with Solid Inspector² revealing issues

> **Speeding up Solid Inspector²**
>
> If you find yourself running Solid Inspector² a lot, which is quite likely depending on your workflow, you may want to assign a shortcut to the command or add the icon to your user interface. Custom shortcuts can be created in the **Preferences** window and the icon can be enabled from **Toolbars** (Windows) or **Tool Palettes** (macOS) from the **Edit** menu.

For this model, three issues prevent it from being a solid group, and they are all the same. While Solid Inspector² can identify other issues with non-solid geometry (this extension version can identify the same issues as the version in SketchUp for Web), most issues seen in imported `.stl` files are the same: surface borders. Since `.stl` files are imported triangular geometries, certain problems are unlikely to exist. Since only faces are being imported, loose edges are not likely to be a problem, and since you are importing a single chunk of geometry, nested instances should not bother you, either.

Let's use Solid Inspector² to get a better look at the holes in our model.

3. Press the *Tab* key.

    This will tell Solid Inspector² to get up close and show you exactly where the errors in the model are. Every time you press *Tab*, Solid Inspector² will move to the next problem.

    The view of the issue that Solid Inspector² chooses to show you may be less than ideal. While it will outline the hole in red and draw a circle around the area, you may start with the camera inside the model unable to tell exactly where the issue lies. If so, orbit around the highlighted area until you can tell what the problem is and where it lies, or toggle **X-Ray** on.

Now that we know what is preventing our imported rhino head from being printable, let's look at a couple of workflows to fix the most common problems with imported geometries.

Repairing existing models

At this point, we know that we have three holes in our mesh, and this rhino head is way too big to fit on the bed of any printer I have seen. While it is always possible that you may find more issues with imported geometry, holes in mesh and models at the wrong size are the most common issues I have seen when working with `.stl` files. Let's run through the steps to address these two issues.

Fixing holes in a mesh

Thanks to Solid Inspector², we know that there are three holes in the mesh and we know exactly where they are. Since the holes are surface borders (holes along the edges of a mesh), Solid Inspector² cannot fix them for us automatically. Fortunately, they are pretty easy to patch.

Up until now, we have run both Solid Inspector and Solid Inspector² in the same way – select a group, then start inspecting. This works perfectly well, and if issues that are discovered can be fixed automatically, this is great! If we run into errors that need to be manually repaired, however, it means stopping Solid Inspector, entering the group, fixing the issues, then exiting and reselecting the group, and starting Solid Inspector again.

This time, instead of running Solid Inspector[2] from outside the group, we will enter the group first before we run it. This will inspect the geometry inside the group and allow us to quickly hop back and forth between inspecting and repairing:

1. Use **Select** to double-click on the rhino head group.
2. Start **Solid Inspector**[2].
3. Press the *Tab* key to zoom in on the first hole in the mesh.
4. Press the *Esc* key to exit **Solid Inspector**[2].

   At this point, we are out of Solid Inspector[2] and in the context of the geometry, zoomed in on the hole. All we have to do at this point is close the hole up. Often, this is as simple as drawing a single edge along the side of the hole. In some cases, however, this hole may need more than a single line to patch it. The easiest way to see how to fill this hole is to see the hidden edges around it. When I repair smoothed meshes, I always turn **Hidden Geometry** on.

5. Toggle **Hidden Geometry** on in the **View** menu.

   At this point, we should have a clear view of the issue, and what we need to do to fix it:

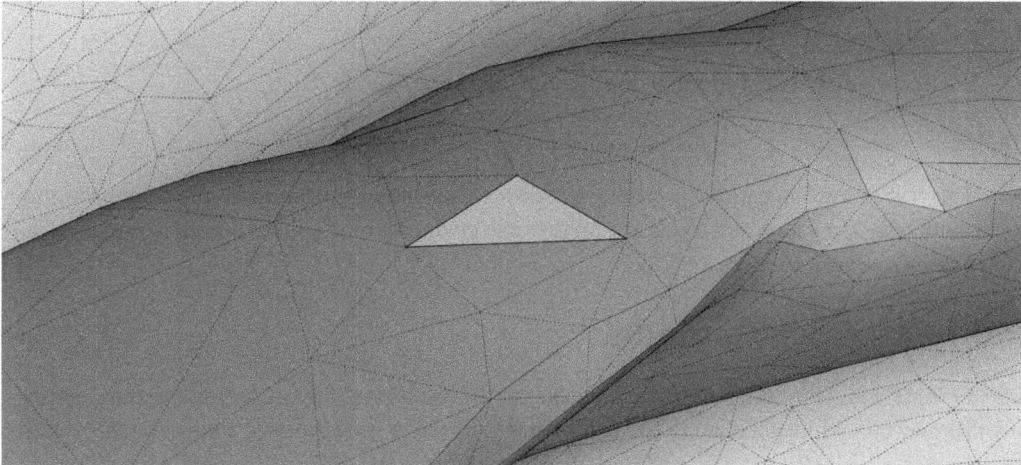

Figure 7.5 – Clearly seen hole in our mesh

6. Use **Line** to draw an edge along one side of the triangular hole to close it.

   If you like, you can use **Eraser** with the **Soften/Smooth** modifier to smooth the edges of the triangle so that it blends in with the rest of the smooth mesh.

7. Start **Solid Inspector**[2] again.

At this point, you should see that the number of errors has dropped to two!

8. Press the *Tab* key to advance to the next error.

9. Repeat *steps 4 through 8* for the next two issues.

    After fixing the third hole, you should be able to run **Solid Inspector**[2] one more time and get a **No Errors – Everything is Shiny** message! At this point, you can exit the group. This is also a great point at which to save your model.

At this point, we now know that our mesh is solid and can be printed. Now, we have to deal with the fact that it is currently larger than my entire 3D printer!

Scaling a model to the correct size

This step is pretty easy, but one that you may need to do regularly when working with other people's 3D models. Since we don't know why this model was created, we don't know what size it should be. We do, however, know what we will use it for, and we know what size we want it to be. While there is no hard rule about the size that a coat hanger needs to be, I will want this to sit far enough from the wall that it can easily hold a coat over its horn, but not so far that anyone will run into it. I think that about 4" should work pretty well. As we did this once already in *Chapter 4, Print-Ready Modeling and Scaling for Export*, I will just give you a quick reminder of the steps:

1. Use **Select** to highlight the group.

2. Use **Scale** with the *Shift* modifier to scale the group so that it is 4" long.

With that, we have a rhino model that we know is solid and the correct size for our intended purpose.

Now, we have a few modifications we need to make before we send anything to the printer. In the next section, let's take a look at making a change to the geometry of the rhino head and designing a mount to which we can connect the head.

Modifying existing models

While there are a lot of models out there, available for download and ready to print, there are just as many that may take a little bit of adjusting before they are perfect. These may be simple changes, such as moving some geometry around or adding a base to a figure, or they may be something big, such as chopping a model in half. Regardless of the modification needed, once you have imported the file and you know that it is solid, you can use SketchUp to modify it as you would any other SketchUp model.

For our rhino head, we are going to make two specific modifications:

*   First, we are going to make the tip of the horn less pointy. Since we are using this to hang clothing on, we do not need a sharp point at the end. While this is something that we could do post-print, we will tackle it in the model so that our hanger is ready to use as soon as it is printed.

- The second modification is to create a wall mount. Our rhino head should fit right into the mount and the mount should be ready to hang on the wall.

Let's start by taking care of that horn.

Using Solid Tools to edit existing geometry

Since we know that our rhino head is a solid object, one of the easiest ways to edit the geometry is using **Solid Tools**. We have done a little bit of editing with **Solid Tools** before, making one piece nest into another. This time, we will create a solid piece specifically to cut away the tip of our rhino's horn.

To do this, let's follow these steps:

1. Off to the side of the rhino head group, use **Rectangle** to draw a rectangle on the ground.

2. Use **Push/Pull** to pull the rectangle up into a box.

3. Use **Select** to triple-click on the box. Then, right-click and choose **Make Group** from the context menu.

    This should give you a simple solid box next to the rhino head group.

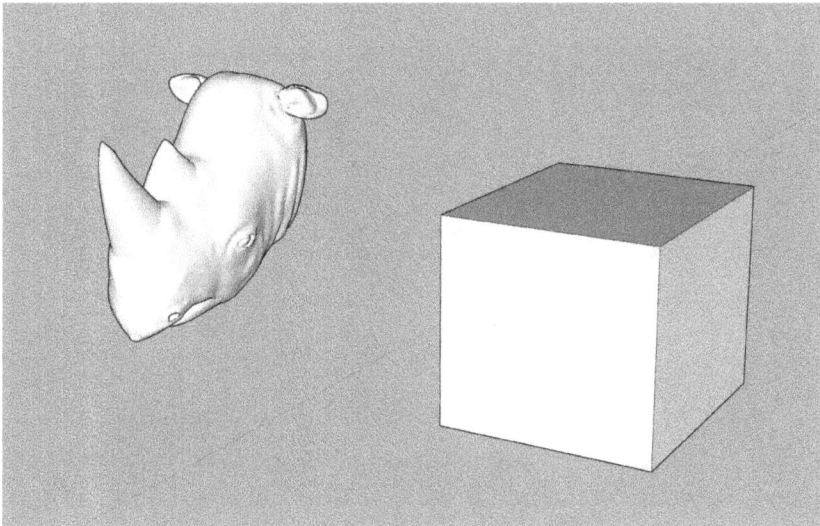

Figure 7.6 – Two solids

The next step is to move the box (our cutter) so that it overlaps with the tip of the rhino's horn.

4. Use **Move** to move the box group so that it covers the tip of the horn.

    You will probably want to orbit around as you move and use the arrow keys to lock inferencing as you go. In the end, you want something similar to what is shown in *Figure 7.7*.

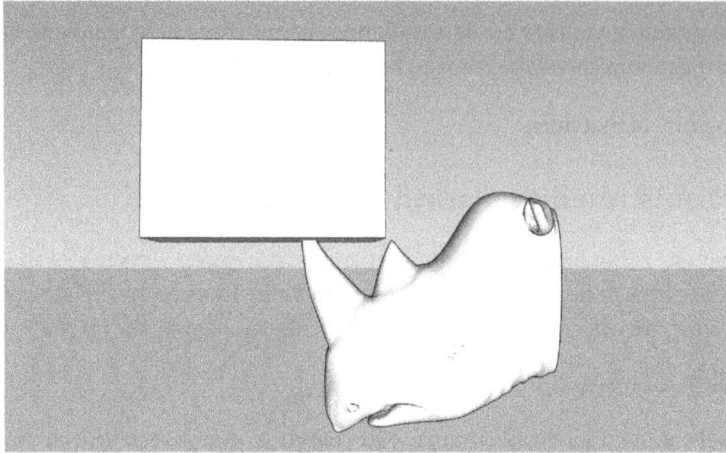

Figure 7.7 – Overlapping solid groups

Just to get the best result, let's also rotate the box so that the cutting face is perpendicular to the horn.

5.   Use **Rotate** and **Move** to turn the box and (if needed) move it so that it looks similar to *Figure 7.8*:

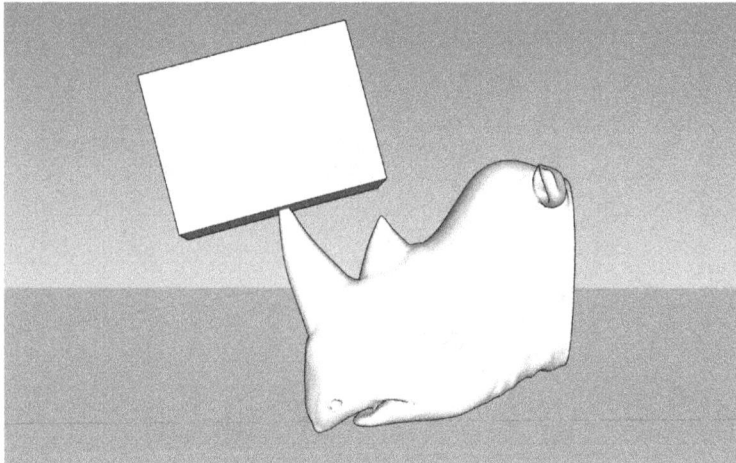

Figure 7.8 – Ready for Solid Tools

This is a great time to save your model. Once that's done, it is time to blunt the horn using **Solid Tools**.

6.   Use **Select** to highlight the box group.

7.   Click on the **Tools** menu, then **Solid Tools**, and then **Subtract**.

8. Click on the rhino head group.

    It will probably take a few seconds to run, but **Subtract** should remove the cube group and the tip of the rhino's horn along with it.

Any time you run **Solid Tools** commands, it is a good idea to check the resulting geometry with Solid Inspector[2]. While the group created as a result of any Boolean operation should be a solid object, there are cases in which the new mesh is missing a small face or something along those lines. The best solution is to keep an eye on **Entity Info** and run **Solid Inspector**[2] regularly.

What if Entity Info and Solid Inspector[2] are not the same?

Every now and then, I will run **Solid Inspector**[2], and it will tell me that the selected group is a solid, while **Entity Info** will say it is just a group. I cannot say why this happens (because I do not know). I do know that the easiest way to fix the issue is to select the group, explode it, and then immediately make it a group again.

Now that our rhino head group is about done, let's model a second group to mount it on.

Using Solid Tools to create connecting parts

We are going to create a new solid – a mount that we can hang on the wall. We will make a hole into which the rhino head can nest. For this to work, we will need to add some geometry to the back of the rhino head. The easiest solution is to use **Push/Pull** to pull the flat section at the back out. To do this, we will need to do a little cleanup first.

Creating a neck

If I rotate around and look at the back for the rhino group, I can see that the flat section at the back is smoothed with the rest of the head in a few places:

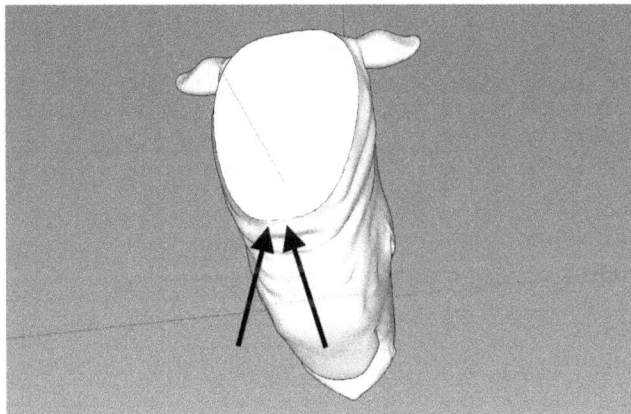

Figure 7.9 – Gaps in the outline on the back of the rhino head

Depending on how much you have smoothed the geometry, you may have more or fewer gaps here or in other places around the head. Let's look at the simplest way to complete this outline so that we can isolate the back face.

Right now, if you enter the group and select any portion of the rhino head, the entire thing will be highlighted. This is because the faces are all smoothed together. To isolate just the back face, we need to make the edges around the face visible:

1. Toggle **Hidden Geometry** on in the **View** menu.

2. Use **Select** to double-click to open the group and then double-click on the back face.

    If we had tried this before with everything smoothed, this would have selected all of the geometry in the head. With the hidden edges visible, the back face can be highlighted, along with all of the edges that define its boundary.

3. Right-click on the face and choose **Soften/Smooth Edges** from the context menu.

4. Toggle both **Smooth Normals** and **Soften Coplanar** off.

5. Toggle **Hidden Geometry** off in the **View** menu.

    Now that the face can be clearly seen and selected, let's pull it out so we have clean geometry to use as a connection.

6. Use **Push/Pull** to pull the back face out by 1/2".

    Right now, our rhino head should look like this:

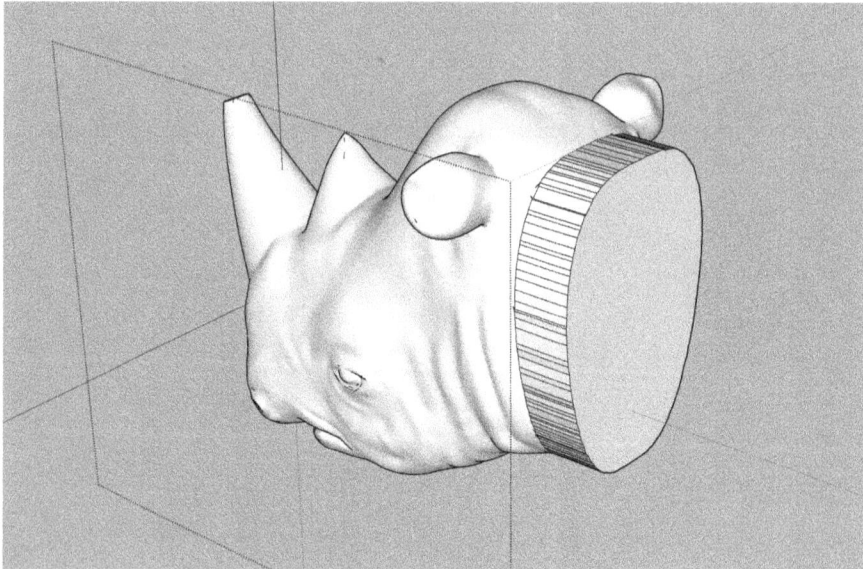

Figure 7.10 – Final rhino head group

Extra smoothing

At this point, you may be tempted to use the **Eraser** or **Soften/Smooth** command to get rid of all those lines on the neckpiece. This is your call. Again, since the `.stl` file that we will soon create will be broken down into even more geometry than we can see, it will not have any effect on what gets output.

The mount that we will create will be a vertical oval and will be slightly deeper than the extension we just added to the neck. This will create geometry that is deep enough that the head will snap into the mount, should be able to hold up to something such as a jacket or hoodie being hung on the horn, and stay connected to the base without any additional glue or mechanical fasteners.

Creating the mount

Let's start by drawing the basic geometry of the mount. It is very important to make sure that you are not in the rhino head group when we create this geometry (as it will end up in a second solid group):

1. Use **Circle** to draw a 96-sided circle with a radius of 2 1/2", starting from roughly the center of the back of the rhino head group:

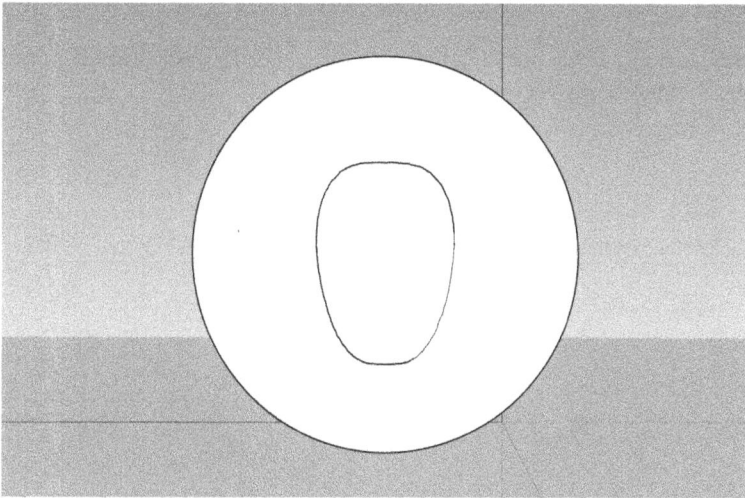

Figure 7.11 – Circle aligned with the back of the rhino head group

2. Use **Select** to highlight the circle.

3. Use **Scale** with the **Toggle Scale About Center** modifier to scale to 80%.

    This should give you a nice oval shape to build the mount from.

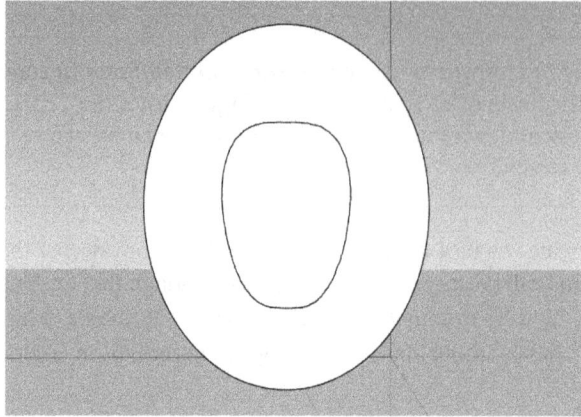

Figure 7.12 – Oval outline for mount

Now, let's add some depth to this flat face.

4.   Use **Push/Pull** to pull the face back (away from the rhino head) by 1/4".

5.   Use **Push/Pull** to pull the front of the oval toward the front of the rhino head by 1/2".

This should result in a piece that is 3/4" thick, 1/2" of which overlaps with the extension on the back of the rhino head:

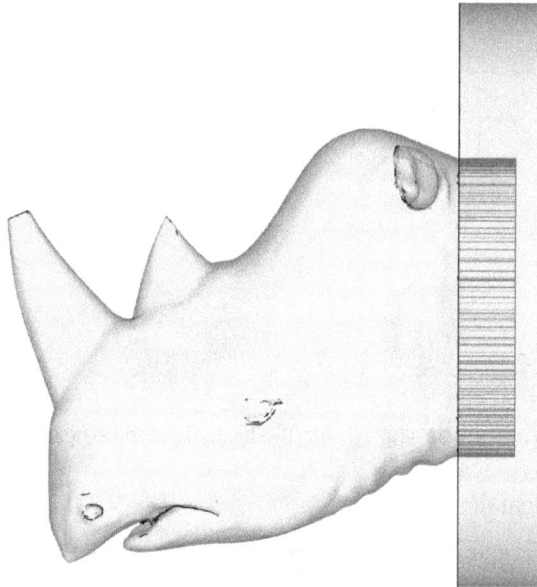

Figure 7.13 – Rhino head and mount (shown with X-Ray enabled)

Let's add a little bit of detail to the mount by adding a quick bevel to the front.

6. Use **Offset** to add a 1/4" offset to the front face of the mount.

7. Use **Move** to push the outside edge of the front face back by 1/4".

   This should create a quick bevel around the front face of the mount.

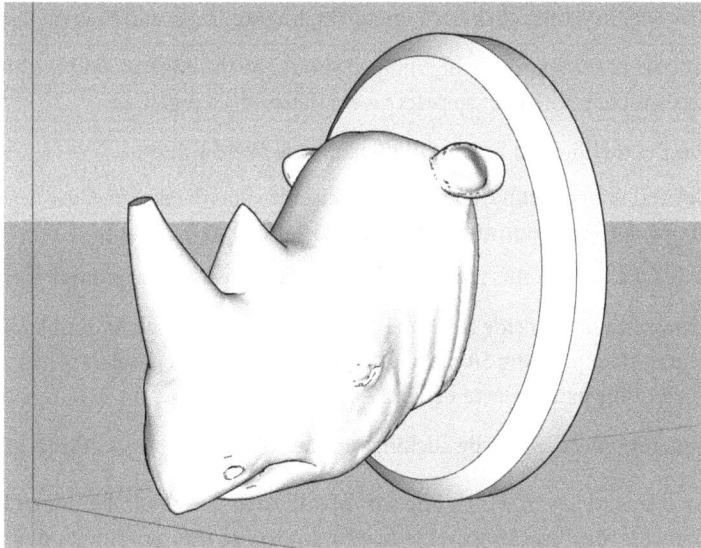

Figure 7.14 – Beveled mount geometry

Now, let's make a hole just slightly bigger than the rhino's neck in the mount.

Subtracting the neck from the mount

Let's group the geometry and use **Solid Tools** to make a hole in the mount:

1. Use **Select** to triple-click on the mount geometry, then right-click, and choose **Make Group** from the context menu.

2. Select the rhino head group, then click on **Tools**, then **Solid Tools**, and then **Trim**.

3. Click on the mount group.

   This will give us a hole in the mount group the exact size of the extension we added to the rhino head. As we learned in *Chapter 6, Modeling Using Solid Tools*, we need to allow extra space between the pieces. Let's hide the rhino head and add a little more space to the hole in the mount.

4. Right-click on the rhino head group and choose **Hide** from the context menu.

5. Double-click to enter the mount group.

   We need to select and offset the outline around the hole in the mount. This can be tricky because it is not a connected shape, but a string of edges. Here is a quick way to select just the outline.

6.   Double-click on the front face of the mount.

This will highlight the front face, as well as the edges around the outside and inside of the face. Now, deselect the face and the outside edge, leaving just the edges around the hole highlighted.

7.   Hold down the *Shift* key, then click again on the highlighted face, and then on the outside edge.

Now, the only part of the geometry highlighted should be the edge around the hole in the face. To keep things simple in the next step, let's weld these edges together.

8.   Right-click on the highlighted edges and then choose **Weld Edges**.

9.   Choose **Offset** and then offset the edges out the distance you determined was best in *Chapter 6, Modeling Using Solid Tools*. For my printer, this distance was 0.15 mm.

10.   Now, use **Push/Pull** to push this new ring down until it meets the bottom of the existing hole.

To clean up the extra lines inside the hole, try using the same trick of double-clicking on the face in the center, then pressing *Shift* and clicking on the face to deselect. Then, you can press *Delete* to get rid of the extra selected edges!

11.   Let's bring our rhino head group by clicking on the **Edit** menu, then **Unhide**, and then **All**.

If we wanted to attach this to a wall with some sort of adhesive, we would be done now and could export our two groups for printing. However, since I want to use this as a hook on which I want to hang my hoodie, I need to add a way to attach it to the wall with a little more strength. I will add a hole to the back of the mount to hang this mount on a screw. To do this, I will model the hole as a solid group and then subtract it from the mount.

You can also do this by modeling the shape shown in *Figure 7.15*, then placing it in the mount group, and using **Solid Tools** to subtract it:

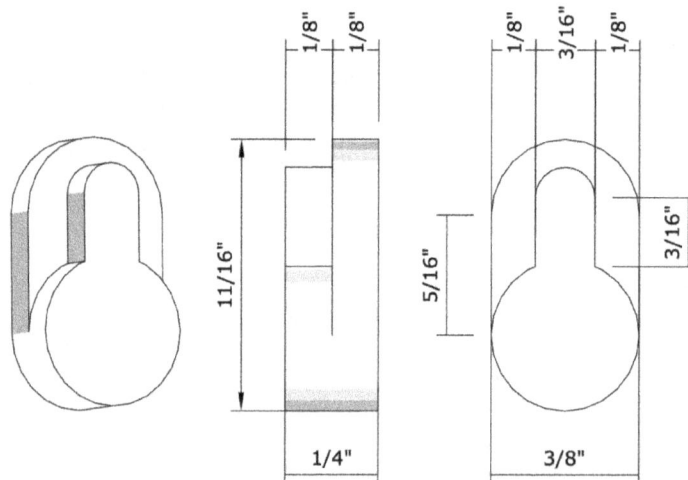

Figure 7.15 – Dimensions for a quick screw hole for the mount

If you do not feel confident modeling this geometry from scratch, you can find a copy of this geometry in the example file specified in the *Technical requirements* section. Use the **Components** panel to drag a copy of the **Screw Hole** component into your model and use that to subtract the space from your mount group.

The result should be a screw hole that will allow you to hang the 3D-printed mount on a standard screw.

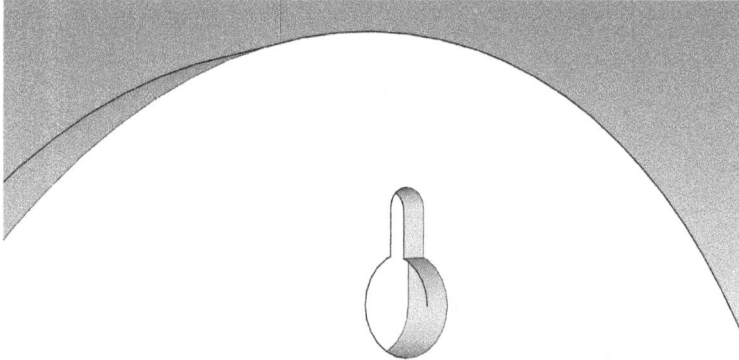

Figure 7.16 – Simple screw hole, ready for mounting

At this point, you are ready to select and export two groups and head into your slicing software!

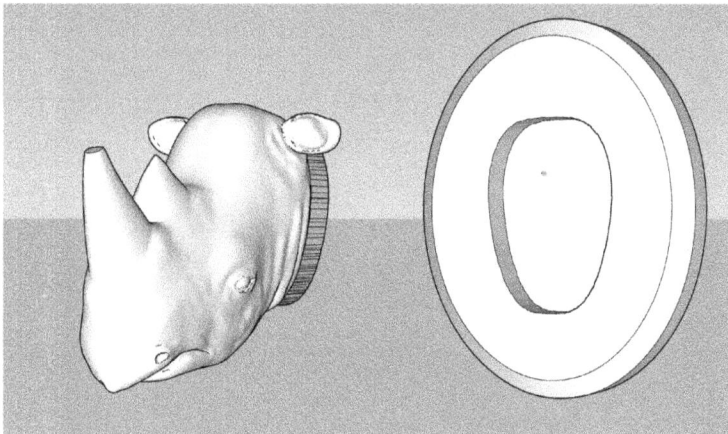

Figure 7.17 – Two solids, ready for export

You did it! You turned a large, holey, imported file into a solid, ready-to-print model, complete with a second piece that is purpose-made for mounting.

Summary

Hopefully, having done everything in this chapter, you not only picked up a tip or two about modeling in SketchUp but also saw how easy it is to work with existing files. We covered the process of repairing and scaling imported geometry, as well as the process of using the native SketchUp tools and **Solid Tools** to modify and add to imported geometry.

At this point, we are ready to put it all together with one final 3D print. Let's tackle that in *Chapter 8, Assembling the Pieces Post-Printing*.

8

Assembling the Pieces Post-Printing

One thing that I love about 3D printing is the ability to help visualize a concept. Yes, I said before that I love being able to create something from nothing and also how amazing it is to have a concept serve a purpose in the real world. Sometimes, however, it is great to make something just so you can see it.

I think that 3D printing has an amazing place in architecture for this reason. For decades, now, architects have made amazing scale models from balsa wood, foam core, and paper to help their customers appreciate their vision well before a plan has been made to start constructing a building. With 3D printing, the process of creating these models goes from days, weeks, or months to minutes, hours, or days.

In this chapter, we will combine some existing model information with some scratch-modeled geometry to help us to visualize a modern hillside home, complete with landscape and trees. We will be modeling the whole thing while being very intentional about the final assembly process and how the pieces will interact after they have been printed.

In this chapter, we will cover these main topics:

- Importing and editing geometry
- Creating new geometry
- Exporting multiple parts for printing

Technical requirements

The steps for this process will all be done using SketchUp for Desktop. You can follow along with most of what is shown in this chapter using the Go or Pro version of SketchUp for Web, but you will have to modify the end (output) of the workflow.

You will also need to download the example file for this chapter, 3D modeling in SketchUp Chapter 8 - Final.skp. This file can be found on 3D Warehouse at https://3dwarehouse. sketchup.com/model/d42f45aa-4738-470d-9f63-9b444e9a0bff/3D-Printing-with-SketchUp-Chapter-8-Final.

Importing and editing geometry

In this chapter, we will be taking a few existing assets and putting them together along with some brand-new geometry. We will be taking existing landscape geometry and placing a model of a house design into it. Then, we will create some trees to place around the landscape to help tie it all together. In the end, we will have an example of what this house may look like once it exists in the real world.

Figure 8.1 – The finished architectural model print

The first thing we will do is take a look at the existing assets. Let's open up the example file for this chapter and see what we are working with.

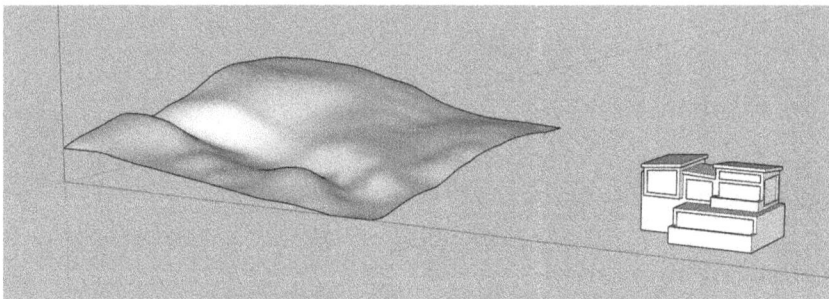

Figure 8.2 – The contents of the example file for Chapter 8

This file contains landscape geometry and a very basic model of a house. If you click on the landscape with **Entity Info** open, you will see that it is a single surface. If you click on the house, you will see that it is a solid group. This means that the landscape is not printable at this point, but the house is. This is good information to know.

If you do a little bit more examining, you will find that the landscape surface is 200 feet wide and the house is 63 feet wide. This puts them in the realm of real-world sizes. This is good to know because we now know that we will, at some point, need to scale these down if they are going to fit on our print bed.

Armed with this information about our model, let's start by making that landscape a printable solid.

Creating a solid landscape

If you are working with any kind of urban design, landscape design, or architecture, you may at some point receive terrain information similar to what is in the example file. The information you receive may show up as a single surface such as the one in this example file, or you may end up receiving a series of points or lines that define the terrain. It is unlikely, however, that you will receive a model that contains a solid group that is immediately printable. Fortunately, there are only a few steps required to turn this single undulating surface into a solid group that we can use for printing.

What do we do with points or lines?

While it is out of the scope of this book, I do want to give you a tip on working with point or line data for the terrain. If you receive this sort of data, you will need to connect the points or lines together with faces in order to make it useful for modeling. **Sandbox Tools** (which is a part of SketchUp for Desktop) has a command called **From Contours** that will stitch together topographical lines to create a mesh. This can work to create a surface that can be used for 3D printing, but it is not an ideal solution.

If you do work with this sort of data with any regularity, I would suggest getting your hands on the TopoShaper extension from the extension developer Fredo6. This extension will create a beautiful, ordered mesh from contour lines or points and even offers you the ability to make it into a solid, automatically. It is available for trial use and purchase through the SketchUcation Plugin Store: `https://sketchucation.com/plugin/716-toposhaper`.

To turn the surface into a solid, we just need to draw some vertical surfaces that run from the sides of the terrain surface down to the ground plane. We can do all of this with the **Line** tool, as shown in *Figure 8.3*.

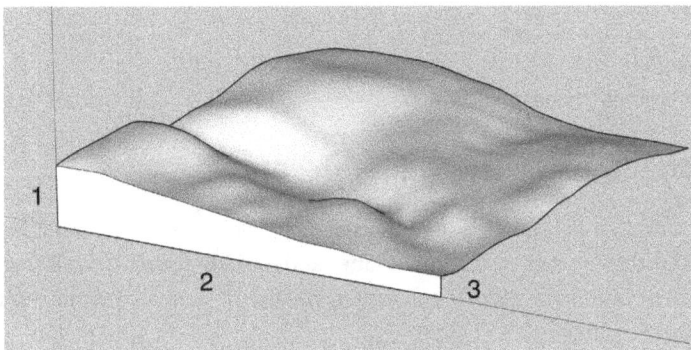

Figure 8.3 – Three edges create the first vertical surface

Follow along to create sides to change the terrain from a single surface into the top of a solid:

1. Start the **Line** tool.

2. Draw the first edge (**1** in *Figure 8.3*), starting at the corner just above the origin and then ending at the origin.

3. Now, draw a second edge (**2**) along the red axis (tap the *right arrow* to constrain movement along the red axis) to right below the lowest corner.

4. Draw a third (**3**) edge up to the corner of the surface to create a vertical face.

   This gives you one side on this surface. Now, use the **Line** command to draw the sides on the other three sides to create a solid geometry.

   Once you have created all four sides, the bottom face will form automatically.

5. Use **Select** to triple-click on the geometry, and then right-click and choose **Make Group** from the context menu.

6. Use **Solid Inspector**[2] to confirm that the group is solid. If not, use **Solid Inspector**[2] to find and repair any problems.

Once complete, you should have a solid group that looks like a square chunk of terrain, cut out of the earth.

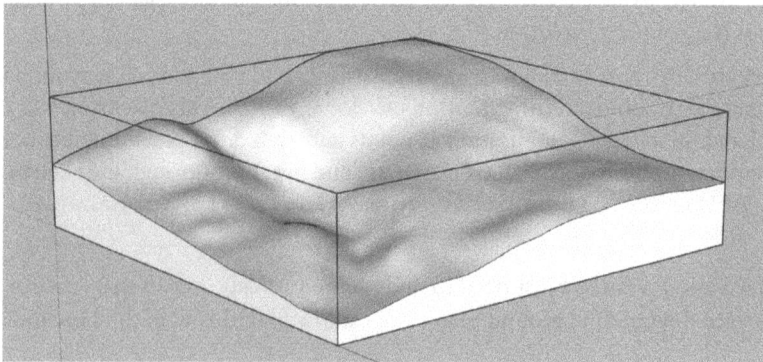

Figure 8.4 – The solid terrain group

Now that we have one solid, we can place the second solid (the house in the example file) into it.

Intersecting the house with the landscape

The house model found in this example file is an extremely basic model of a modern house. It was intentionally created with exaggerated window shapes and limited detail, in the knowledge that the final print will be just over an inch wide. If you prefer, you are, of course, welcome to add details or substitute it with your own house model.

If you do choose to use another file, make sure that you take the following design features into consideration:

- **Single flat bottom**: In the example model, notice that the bottom sits flat on the ground. This is intentional, as this model will sit in a hole in the landscape but still be removable. When removed, the single, flat base means it will sit just as well on my desk as it does in the terrain print.

- **Vertical sides**: If I had modeled this building as it will be built, I would likely have included footers or other foundation details that would expand the footprint of the house. Since I want this model to easily slide in and out of the terrain print, the vertical sides from the bottom are intentionally simple.

- **Tall foundation**: Speaking of the foundation, you may have noticed that there is about 6' of extra geometry below the "floor" of the house. This is again intentional, as I wanted to have a section of the structure that could drop into the terrain without losing any detail below the ground.

- **Simple details**: Just like the house we worked on back in *Chapter 4*, *Print-Ready Modeling and Scaling for Export*, the details of this house are simplified, with smaller details removed, and those that are present are exaggerated. To be sure, if this were an initial massing model, considering how it might sit into a chunk of terrain, you may not have a lot of details to use anyway.

Now that we know why this model looks like it does, let's get it dropped into the terrain.

Arranging and scaling

For this step, we will only place the house into the terrain, and we will not cut a hole into the terrain group until after we have scaled the model down. Let's start by rotating the house and then moving it so that it sits into the terrain:

1. Use the **Rotate** tool to turn the house 45 degrees clockwise.
2. Use the **Move** tool to grab the house by the front corner (the point called out in *Figure 8.5*) and drag it onto the terrain.

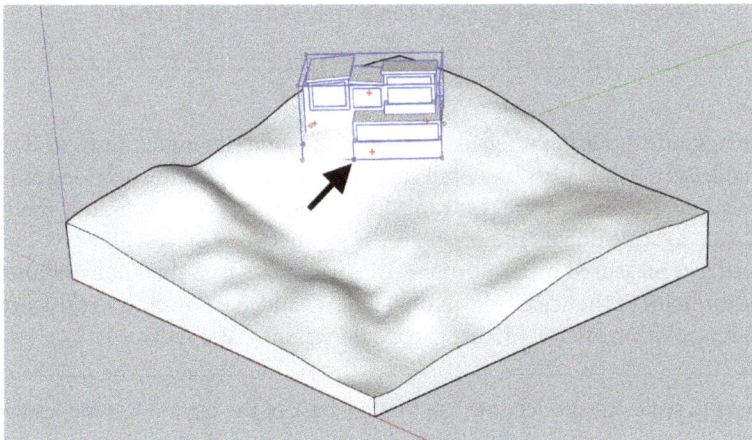

Figure 8.5 – The house preliminarily dropped onto the terrain

Moving selected geometry over a mesh like the one in our terrain group can be tricky. Remember that the point that you are using to move the group will snap to the points that make up the terrain mesh. As you move that point along the undulating surface, the rest of the house will disappear and reappear above the terrain. Use your initial move to get the house approximately where you want it and then fine-tune from there. Once you have the house in a good spot up on the hill, use **Move** again to drop it down so that the bottom of the house group is completely below the terrain surface.

3. While still in **Move**, tap the *up arrow* to lock to the blue axis and slide the house down below the surface of the terrain group.

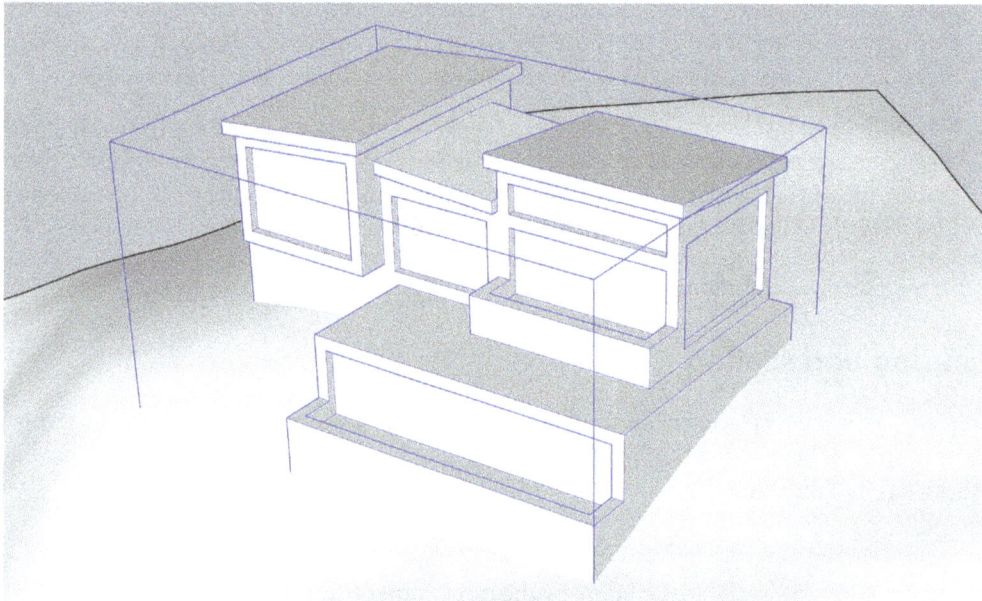

Figure 8.6 – The house group completely in the terrain group

When to scale?

There is not a single rule that states when to scale your real-world size model down to 3D printing size. I try to model as much as possible at full size and put off scaling down as long as possible. In the case of this model, once we are ready to add holes for one group to fit into another, that is the perfect time to scale down. The reason is the gap between pieces. Since we know that the gap we need was measured in 3D printing size increments (less than one millimeter), we want to add these holes to the scaled-down model. If we added 0.2 mm to a real-world-sized hole and then scaled down afterward, that gap would virtually disappear.

We are going to scale our model down so that the base of the terrain is 4 inches wide.

4. Use the **Scale** tool to draw from one corner of the terrain group to the next.

    Since the base of the terrain group is a square, it does not matter which side we draw along.

5. Type 4 ", and then press the *Enter* key.

    This will prompt SketchUp to ask whether you want to rescale the entire model.

6. Click the **Yes** button.

7. Click **Zoom Extents**.

With that, we have our model at a print-ready scale.

Building a cutter

Now, we need to punch a hole in the terrain group that will accommodate the house print. One option at this point would be to use the house (which is a solid group) and the **Trim** Solid Tool to cut a hole in the terrain. However, since we want this house to fit into the hole and be easily removed, we need to make the hole bigger. In my case, I will choose to make the hole 0.2 mm larger than the house so that I can easily pick it out of the terrain any time I like.

Instead, I will make a brand-new piece that I can use just to cut the larger hole. I will do this by copying the bottom of the house into a new group and then making it a bit larger. To do this, follow these steps:

1. Let's temporarily hide the terrain group by using **Select** to highlight the terrain group, right-clicking, and then choosing **Hide** from the context menu.

2. Double-click into the house group and then orbit so that you can see the underside of the house.

3. Highlight the bottom face of the house, click on the **Edit** menu, and then click **Copy**.

4. Click away from the house to close the house group.

5. Click back on the **Edit** menu and choose **Paste in Place**.

    This will place an exact copy for the face of the bottom of the house group outside of the existing house group. Let's put this face in its own group, make it 0.2 mm larger, and then make it 3D.

6. Right-click on the newly pasted face and choose **Make Group** from the context menu.

7. Double-click on the new face to open the group.

Hiding the rest of the model

Generally speaking, I keep **View | Component Edit | Hide Rest of Model** toggled on when I do most of my modeling. With this option turned on, everything disappears when I enter a group to edit it. In the case of what we are walking through right now, this means that once I double-click on the face, the existing house model goes away. This makes it much easier to do the editing I need to do without the house getting in the way.

We could use **Push/Pull** right now to pull this face up into 3D space and make a solid, and then go back in and pull each side out the extra 0.2 mm. Now, that is a lot of work and is prone to error (I feel that I would be very likely to miss a face or accidentally pull one out twice). Instead, we will use **Offset** to grow the whole base by 0.2 mm and then pull the face up afterward.

8. Use the **Offset** tool to add a 0.2 mm offset to the face.

    This should give you a thin outline around your original face.

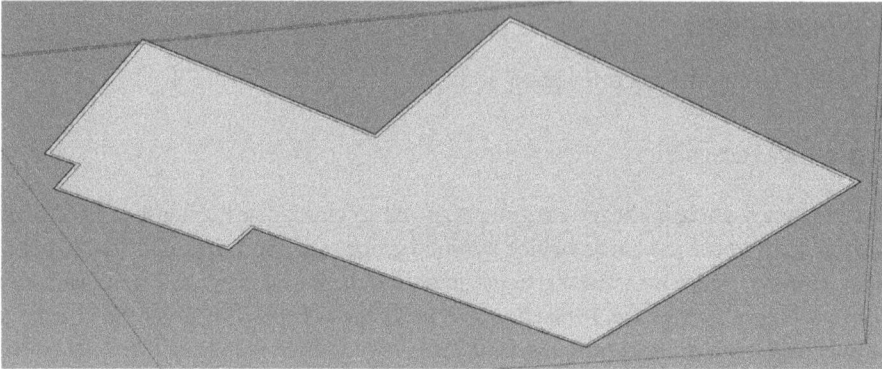

Figure 8.7 – A 0.2 mm face offset

Before we can pull this face into 3D space, we need to get rid of the edges from the original face. The easiest way to do this is using the same select trick we used on the rhino head mount in *Chapter 7, Importing and Modifying Existing 3D Models*.

9. Use **Select** to double-click the inside face, and then hold down *Shift* and click on the face again.

    This will leave only the edges from the original face highlighted.

10. Press the *Delete* key.

    We are now ready to take this face into 3D space.

11. Use **Push/Pull** to pull the face up.

    I don't have an exact amount to pull the face up by; just make sure that it comes up so that it will extend past the terrain group's surface. I pulled my face up taller than the original house for good measure.

12. Use **Select** to click outside the new group.

This will close the group and bring the house group back. This is a great time to highlight the new group and verify that it is a proper solid.

Cutting a hole in the terrain

Now that we have all the pieces we need to cut a hole out of the terrain solid, we will follow these steps to move ahead:

1. Let's get our terrain group back by clicking the **Edit** menu, then **Unhide**, and then **All**.

    With that, you should have the new cutter group covering most of your house and popping out of your terrain.

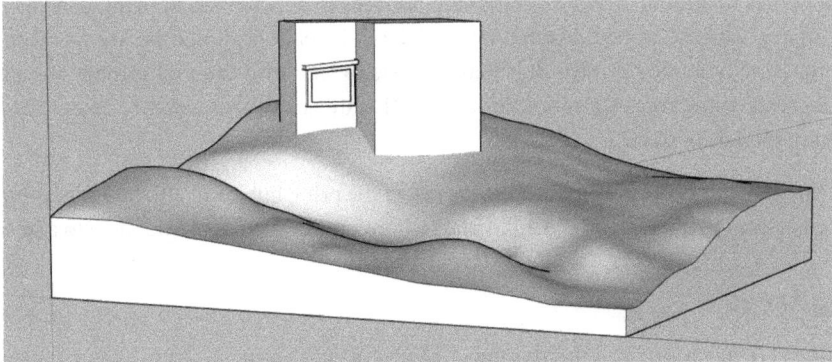

Figure 8.8 – Three groups, ready for Solid Tools

2. Use **Select** to highlight the new cutter group.
3. Start the **Subtract** Solid Tool.
4. Pick the terrain group.
5. Verify that the terrain group is still solid.

    With that, you should have a hole in your terrain with plenty of space for your model to swell slightly and still fit in.

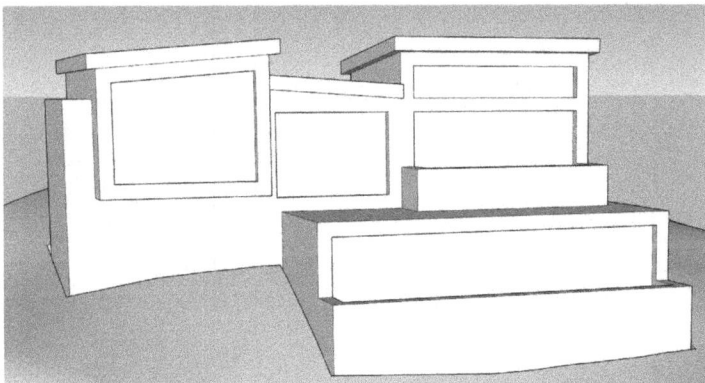

Figure 8.9 – The house group sitting in the terrain group

These two groups look great! Now, it is time to create some new geometry from scratch to add to our print. Let's make some trees!

Creating new geometry

While our two groups could be printed and make a really nice little visual, showing how the house would sit in this terrain sample, I think that we can push a little further and add some trees to the print.

Now, when it comes to adding complex items such as trees to a rather small print, such as the one we are working on, we have to decide what level of detail is realistic. Since we are working with a preliminary massing type model, I think it makes sense to lean into the preliminary visualization aesthetic and create some trees that look more like 2D cutouts of trees, rather than try to make realistic-looking models of trees.

Not only will this fit better in the look and feel of the rest of the print, but it will also be easier geometry to print, as we can orient the trees flat on the build plate without needing to figure out how to add support.

Modeling trees

Rather than trying to detail out, step by step, how to make the exact tree model I created, I will present you with some basic dimensions and give you a few tips for modeling this sort of geometry. First, here is the tree that I ended up creating with the important dimensions:

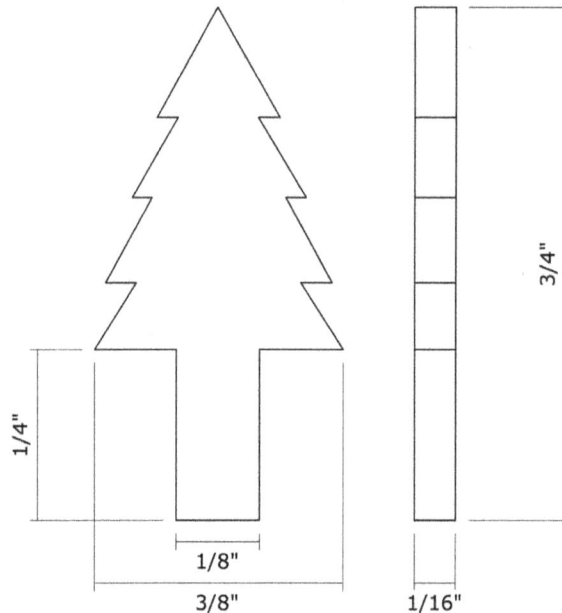

Figure 8.10 – The basic tree dimensions

With this basic set of dimensions, you should be able to create a quick tree model. Before diving in, though, I will present a few tips that will make this modeling process as easy as possible:

- **Model flat**: When modeling what is essentially a 2D shape, it may be easiest to draw a large rectangle on the ground and then snap the camera to the **Top** view. Then, when drawing edges, make sure that your cursor snaps to the face, thus keeping all your points on one plane.

- **No length snapping**: This is especially important when modeling something as small as this tree. Go into the **Windows** menu, and then click **Model Info**. Select the **Units** tab and make sure that **Enable length snapping** is turned off. If this is on, it will be very tricky to draw in the branches of the tree, which is less than an inch tall.

- **Model half**: In order to make your tree symmetrical, model one half of the tree, and then use **Rotate** to copy the tree half.

- **Sketch, move, and clean up after**: When you are making a shape like these trees, don't stress about nailing a perfect outline on your first go. Feel free to add lines, cross them over and trim them back, or move edges around. In the end, all that matters is the outline, and you can always use **Eraser** to go back and clean up the extra edges.

Keeping this in mind, here is what my modeling process looked like.

Figure 8.11 – The steps my geometry went through to become a tree

Once you have the final geometry, ensure to make it a group and verify that it is a solid.

Now that we have a tree model, we just need some holes in the terrain to stick them in once we have printed everything.

Intersecting trees with the landscape

Much like the process we used to put a hole in the terrain for the house, we will be using a cutter group (geometry that only exists to punch a hole in another group) to put holes into the terrain for our trees to stick into. Let's model a simple box that represents the exact opening size we need to snap the trees into, and then use it to subtract some holes from our terrain group:

1. Using the **Rectangle** or **Line** tool, draw a rectangle that is the size of the base of our tree (1/8" x 1/16").

2. Use **Offset** to offset this outline just enough so that there will be a tight fit between the tree and this hole (for my model, based on the connection test from *Chapter 6, Modeling Using Solid Tools*, I know that I need a 0.15 mm additional gap to create a good solid friction fit connection).

3. Remove the extra edges from the rectangle, use **Push/Pull** to pull it up 1/2", and make it into a group. Verify that it is solid.

    I pulled the rectangle up twice the height of the trunk on our tree model intentionally. We will see why that is important when we start placing this cutter into the uneven terrain in *step 6*.

    Before we can start cutting this shape out of the terrain, we need to turn the group. I chose to angle the trees the same as the house, so someone looking directly at the front of the house would also see the front of every tree.

4. Use **Rotate** to turn the cutter group 45 degrees.

5. Use **Select** to highlight our cutting group.

6. Start the **Move** command and pick up the cutter group by inferencing the midpoint of one of the edges of the box, as shown in *Figure 8.12*:

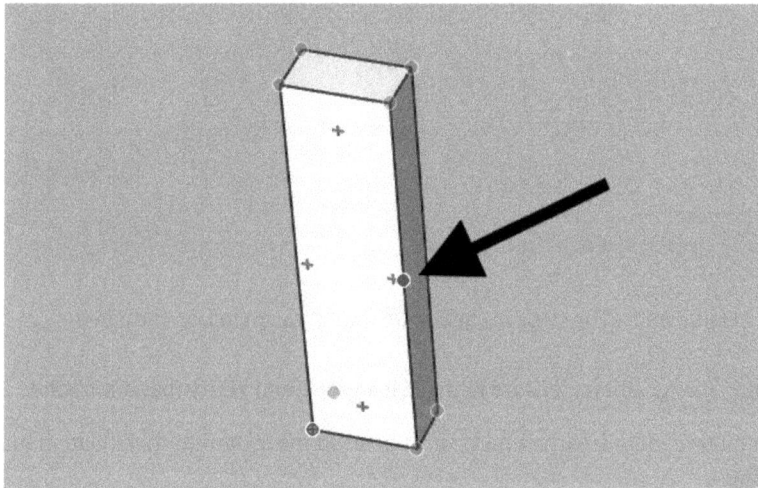

Figure 8.12 – Pick the cutter group from this point

7. Move the group onto the terrain and place it downhill and to the right of the house, as shown in *Figure 8.13*:

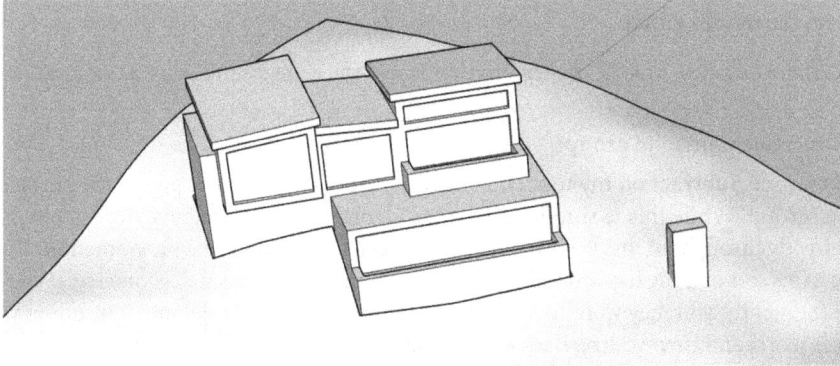

Figure 8.13 – The first cutter group on the terrain group

By dragging this group by the middle of a vertical edge and snapping to a point on the terrain, we are assured that we will create a hole that the trunk can fit into. Additionally, by having the cutter twice as tall as the hole we need, we are assured that the cutter will always extend past the top of the terrain geometry.

8. Use **Move** along with the **Stamp** modifier to place five or six copies of the cutter on either side of the model, as shown in *Figure 8.14*:

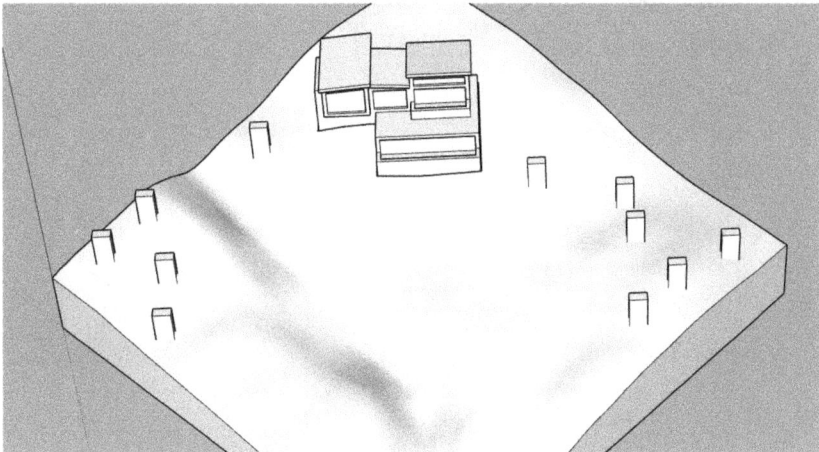

Figure 8.14 – 11 copies of the cutter in the terrain

Now, use **Subtract** to punch a bunk of holes in the terrain group.

9. Use **Select** to highlight one cutter group.

10. Activate **Subtract**.

11. Click on the terrain group.

This will create a new hole in the terrain that is just the right size to snap a tree into.

Solid Tools create non-solid groups

When I first used **Subtract** on my model, I ended up with a rectangular gap in the surface of my terrain group. While this is simple enough to identify and fix (we have done a few fixes like this already throughout this book), I did want to call attention to this phenomenon. In my experience, this sort of thing happens when you are working with very small geometry. If you see this issue frequently, you may want to consider scaling your entire model up 100x, doing your Solid Tools work, and then scaling back down afterward. As far as I could tell, on this model, there was only one issue on the first subtract, and then everything worked perfectly afterward.

---

12. Use **Select** to highlight the next cutter, and then subtract it from the terrain group.

13. Repeat for the remaining cutter groups.

This will give you a series of matching holes across your terrain group, as shown in *Figure 8.15*:

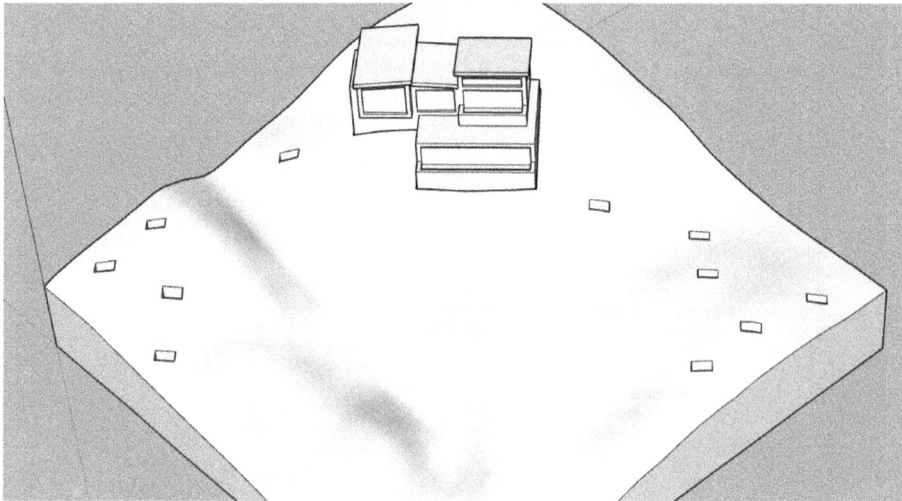

Figure 8.15 – 11 holes ready for 11 trees

With this, all of the groups we need are done, and we are ready to prepare the file for the printer. Of course, if you wanted to, you could rotate your tree model and place copies into each of the holes in your terrain group, although this would be done just so you could see what the model looks like in the end and does not prepare you for printing it. In fact, placing copies of the tree group throughout the terrain creates a geometry that you will want to delete before exporting.

Sometimes, though, we do things in SketchUp just because they look cool. For a case in point, check out *Figure 8.16*!

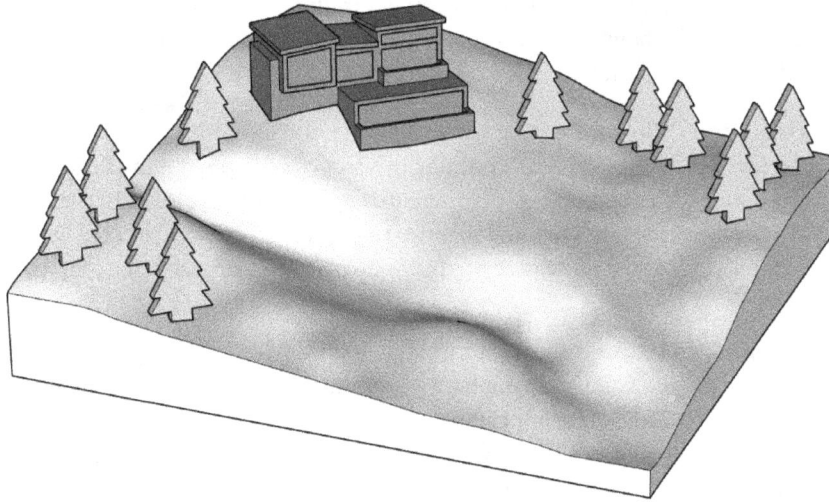

Figure 8.16 – The final pieces, assembled in SketchUp

Now that I got that out of my system, let's get these groups prepped for export and then put them all together.

Exporting multiple parts for printing

We exported more than one piece in the previous chapter (*Chapter 7, Importing and Modifying Existing 3D Models*), but in this chapter, I want to get into more of the *hows* and *whys*, especially in relation to a model like the one we just made that has 13 pieces to be printed.

Prepping multiple groups for output

There are two ways you could approach exporting multiple groups for printing. You can either arrange the pieces exactly as you want to print them and export them as a single file, or you can select and export each group independently, arranging them for printing outside of SketchUp.

I will always recommend exporting each piece individually, rather than as a group. This gives you more control once you leave SketchUp and allows you to print each piece separately. Looking at our model, I want to print the house on my SLA printer, and my landscape in white filament and my tree in green on my FDM printer. The easiest way to do this is to export three different files.

Of course, if you are using Sketch Up for Web, this means copying and pasting each group into its own model, but in SketchUp for Desktop, it is as simple as selecting and exporting each group. A little extra work will now give you much more flexibility.

In our model, it is pretty straightforward to select the terrain group and export it as a .stl file, and then do the same for the house model. The trees, on the other hand, deserve a few moments of consideration before exporting them, as follows:

- **Print orientation**: When exporting something like the tree, I like to try to put it in the orientation that it will print in. When printing this sort of geometry on my FDM printer, I lie it flat on the ground, so that is how I will orient the tree group before exporting it. This is why, in the *Creating new geometry* section, I mentioned that placing copies into the holes was not really a step forward, as the initial tree on the group plane was probably the only one that we would export.

- **Consider copies**: The other thing to think about is how many of these groups to export. In some cases, you may be tempted to duplicate the geometry (remember, we will need 11 trees in total) and export a single file for printing. I will always export a single group in cases such as this one. Exporting a single group gives me more control when preparing for printing. If all 11 trees are laid out in a single file, I cannot move them around once I get to my slicer. If the arrangement I made in SketchUp does not fit on my print bed, I am out of luck and have to head back to SketchUp to fix it. If I export one tree, I can use my slicer to make all the copies I want and move them around the print bed without any rework.

- **Minimizing supports**: Additionally, the support work that is done in the slicer is something to think about. While this tree in our model will not require any support before printing, other models that need to be multiplied may. In these cases, it is best to export one instance of the model and import it into your slicer, where you can add and fine-tune supports. Then, once the model is perfectly supported and ready for export, make copies. Exporting all copies together could mean a lot more work for creating supports.

The most important thing to think about as you export these groups is doing the right work in the right program. SketchUp is the best tool for creating and editing geometry. Your slicer is likely the best tool for orienting and copying geometry for printing. With this mindset, it is pretty easy to figure out how to export each solid group.

Once you have done so and printed all the pieces, it's time to put it all together!

Assembling after printing

While most of this is pretty self-explanatory, I wanted to call attention to a few specific things when putting the printed pieces together:

- Since we designed the hole that the house fits into larger than we needed (I used a 0.2 mm gap), the house (which I printed on my SLA printer) literally drops right into the hole and can be lifted out just as easily. If, at some point in the future, I decide that I do not want it to come loose, I could easily put a drop of glue into the hole and press the house in, permanently.

- As for the trees, the 0.15 mm gap means that I was able to snap each tree into a hole and have it lock into place. There was no need for any adhesive at all! Despite that, I am able to firmly grasp a tree and pull it free from the terrain if I need to.

- This is a great example of how pieces can interact differently based on a very small change in the gap between pieces. A 0.05 mm change in the gap means a loose-fitting piece or a friction fit that will keep a piece in place indefinitely.

This not only ties all the pieces of this model together but also the concepts we have explored throughout this entire book!

Summary

In this chapter, we reviewed many of the concepts that we covered in the previous seven chapters and even learned about a few new ones. We turned a 3D landscape import into a printable solid and covered the use of dedicated cutter solids. We dove deeper into the 3D printing piece gap and fine-tuned our process for exporting multiple pieces as `.stl` files.

As we wrap up this book, I sincerely hope that you have enjoyed reading it and have learned something new. I love modeling in SketchUp as much as I love 3D printing, so the marriage of the two makes so much sense to me. I hope that you now find yourself armed with new tips, tricks, and workflows that will allow you to spread your enthusiasm and create something amazing.

Index

Hi!

I am Aaron Dietzen, author of 3D Printing with SketchUp Second Edition. I really hope you enjoyed reading this book and found it useful for increasing your productivity and efficiency in 3D Printing with SketchUp.

It would really help me (and other potential readers!) if you could leave a review on Amazon sharing your thoughts on 3D Printing with SketchUp Second Edition here.

Go to the link below or scan the QR code to leave your review:

`https://packt.link/r/180323735X`

Your review will help me to understand what's worked well in this book, and what could be improved upon for future editions, so it really is appreciated.

Best Wishes,

Aaron Dietzen

‹packt›

www.packtpub.com

Subscribe to our online digital library for full access to over 7,000 books and videos, as well as industry leading tools to help you plan your personal development and advance your career. For more information, please visit our website.

Why subscribe?

- Spend less time learning and more time coding with practical eBooks and Videos from over 4,000 industry professionals

- Improve your learning with Skill Plans built especially for you

- Get a free eBook or video every month

- Fully searchable for easy access to vital information

- Copy and paste, print, and bookmark content

Did you know that Packt offers eBook versions of every book published, with PDF and ePub files available? You can upgrade to the eBook version at packtpub.com and as a print book customer, you are entitled to a discount on the eBook copy. Get in touch with us at customercare@packtpub.com for more details.

At www.packtpub.com, you can also read a collection of free technical articles, sign up for a range of free newsletters, and receive exclusive discounts and offers on Packt books and eBooks.

Other Books You May Enjoy

If you enjoyed this book, you may be interested in these other books by Packt:

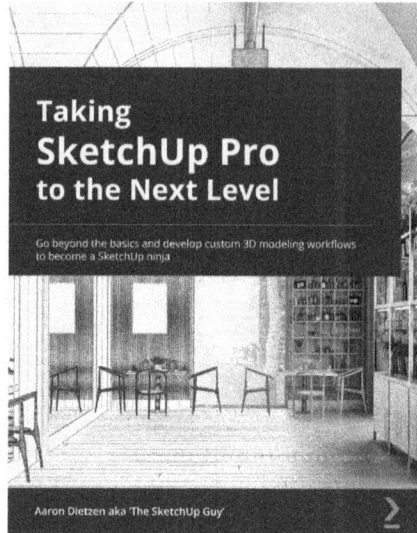

Taking SketchUp Pro to the Next Level

Aaron Dietzen

ISBN: Aaron Dietzen

- Recap the basics of navigation and SketchUp's native modeling tools
- Modify commands, toolbars, and shortcuts to improve your modeling efficiency
- Use default templates, as well as create custom templates
- Organize your models with groups, components, tags, and scenes
- Analyze your own modeling workflow and understand how to improve it
- Discover extensions and online repositories that unlock the advanced capabilities of SketchUp
- Leverage your existing SketchUp Pro subscription for even better results

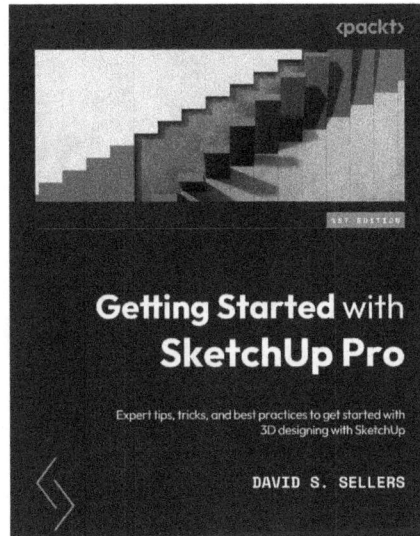

Getting Started with SketchUp Pro

David S. Sellers

ISBN: 978-1-78980-018-0

- Design massing 3D models and preliminary design

- Determine efficient and productive ways of working with SketchUp

- Explore various SketchUp tools and understand the complete functionality of this software program

- Create and edit components and explore component options

- Explore SketchUp extensions, 3D Warehouse, and additional tools and resources

- Get a complete walkthrough of editing tools, materials, and components in SketchUp

Packt is searching for authors like you

If you're interested in becoming an author for Packt, please visit authors.packtpub.com and apply today. We have worked with thousands of developers and tech professionals, just like you, to help them share their insight with the global tech community. You can make a general application, apply for a specific hot topic that we are recruiting an author for, or submit your own idea.

Download a free PDF copy of this book

Thanks for purchasing this book!

Do you like to read on the go but are unable to carry your print books everywhere? Is your eBook purchase not compatible with the device of your choice?

Don't worry, now with every Packt book you get a DRM-free PDF version of that book at no cost.

Read anywhere, any place, on any device. Search, copy, and paste code from your favorite technical books directly into your application.

The perks don't stop there, you can get exclusive access to discounts, newsletters, and great free content in your inbox daily

Follow these simple steps to get the benefits:

1. Scan the QR code or visit the link below

    https://packt.link/free-ebook/9781803237350

2. Submit your proof of purchase
3. That's it! We'll send your free PDF and other benefits to your email directly

www.ingramcontent.com/pod-product-compliance
Lightning Source LLC
Chambersburg PA
CBHW080554220326
41599CB00032B/6474